BUYING PRODUCE

BUYING PRODUCE

A Greengrocer's
Guide to Selecting
and Storing
Fresh Fruits
and Vegetables

by Jack Murdich

HEARST BOOKS · NEW YORK

Library of Congress Cataloging-in-Publication Data

Murdich, Jack.
Buying produce.

Includes index.
1. Fruit.   2. Vegetables.   3. Marketing (Home
economics)   I. Title.
TX397.M87   1986   641.3′4   85-30192
ISBN 0-688-05959-7 (pbk.)

Printed in the United States of America

First Edition

1 2 3 4 5 6 7 8 9 10

BOOK DESIGN BY LINEY LI

To my wife, Florence, who is a peach

My daughters, Peggy and Kathy, a great "pear"

and my grandson, Jimmy Hoover,
the apple of my eye

Acknowledgments

Thanks to the United Fresh Fruit and Vegetable Association; the Market News Service of the U.S. Department of Agriculture; the national trade newspapers of the produce industry, *The Packer* and the *Produce News*; D'Arrigo Brothers Co. of New York for the generous use of their office at the Hunts Point Market; Varian Cassat, food editor of the Gannett Westchester Rockland newspapers; and to Joan Nagy, my editor at Hearst Books.

Contents

VEGETABLES AND HERBS

BUYING PRODUCE

Introduction

The produce section is by far the most difficult area in the supermarket in which to shop wisely. Part of the problem is that when shopping for fresh fruits and vegetables there are no brand names or labels to guide you. When buying processed foods or dry groceries, you can comparison shop until you find a brand that suits your taste and pocketbook, and for the next umpteen years, all you have to do is repurchase the same label. When you shop for fresh fruits and vegetables, every day is a new ball game. Your favorite market may have superb strawberries on Monday, but when you return on Thursday to duplicate your purchase, the berries may be awful.

When purchasing almost any other commodity, the price tag is the key to quality. The most expensive seat in the theater affords the best view. A Mercedes gives you a better ride than a Volkswagen. A $150 pair of shoes will have better material, workmanship, and style than a pair that costs $40. Yet the pint box of strawberries that sells for a modest 69¢ during the peak of the season in May is nearly always superior to the strawberries that sell for $5 to $6 per pint in December. Yet a low price tag is no guarantee of quality or value; price without quality is never a bargain. Overripe or grass-green bananas selling at 29¢ per pound aren't nearly as good a buy as fine ones selling at 59¢ per pound. Yet sometimes the lower-priced bananas are of better quality than the ones with the higher price tags.

Each year the prices of fresh fruits and vegetables climb higher than the previous year, while at the same time the quality shows slippage. At the produce counter of an average supermarket the quality of about 30 percent of the "fresh" fruits and vegetables leaves much to be desired. Unwise purchases made at the produce counter at best will lack flavor; at worst they will wind up in the garbage can.

The aim of this book is to help the consumer get a fairer shake at the produce counter by spelling out how to identify top quality and, more important, how to recognize and avoid the mediocre. If you follow these guidelines, you will have eaten your last mushy apple, flavorless melon, or sour grape.

SEASONS

"To everything there is a season." This quote from the Bible is no longer accurate when it pertains to fresh produce. Seasons for fresh fruits and vegetables have been greatly extended in the past forty years, thanks to modern refrigeration, speedier transport, and the development of new varieties. Items that used to be available eight to twelve weeks of the year are now offered for sale eight to twelve months of the year.

New varieties, some of which ripen earlier or later than their predecessors and have much longer shelf lives, are the result of the combined efforts of the growers, agricultural schools, and Department of Agriculture experimental stations. Each year countless new strains of vegetables and fruits are perfected. Forty years ago we had only two varieties of nectarines; today there are more than one hundred varieties, and there are still more to come. This combination of growers, colleges, and USDA is also responsible for great strides in the improvement of agricultural know-how, one of the most important being the perfection of controlled atmosphere (CA) storage, which has more than doubled the shelf life of apples and has extended the availability of grapes and pears.

Modern refrigeration coupled with larger and speedier transport has completely changed the produce business. Not too long ago we were still using horses and wagons and primitive trucks filled with cakes of ice to haul produce. Today we have eighteen-wheeled, refrigerated trailers that speed produce via eight-lane super highways. These vehicles can span the country in about five days. It takes less than three days to haul produce from Florida to New York.

The speed of these trucks is a snail's pace when compared to the speed of the cargo jets that now fly produce from any part of the nation, as well as from the four corners of the earth, to our metropolitan wholesale markets. In the bleakest midwinter, consumers can purchase almost any item of fresh produce. Even if there is snow on the ground, retail markets in our northern cities offer strawberries, raspberries, blueberries, melons, vine-ripe tomatoes, endive, pineapples, papayas, mangoes, artichokes, and asparagus that are flown in from New Zealand, Australia, Hawaii, Chile, Central America, Africa, Europe, and Israel. We even fly in fresh ginger root from the Fiji Islands. In Grandma's day, if you lived north of the Mason-Dixon Line, fresh strawberries arrived late in spring and wound up around the Fourth of July. Today strawberries are available twelve months of the year. Cargo jets are used not only to fly in the more exotic and costly items but to compensate for temporary shortages. When there is a hot market (short supplies and high prices), items like lettuce, cauliflower, peas, and beans are brought in via jet.

Senior citizens can still remember when supplies of fresh fruits and vegetables were very limited during the winter and early spring. Back in the twenties and thirties, by the time February rolled around the fresh produce cupboard was bare. The grocery stores of that era (supermarkets had yet to make their debut) sold very little other than old cabbage and some root vegetables during the off-season. Cabbage, potatoes, onions, beets, carrots, and turnips were locally grown, harvested prior to the first frost, and stored in barns and root cellars for use in the winter. In those days the California produce industry, still in its infancy, was hampered by inadequate refrigeration and sluggish transportation. More often than not, by the time rail cars carrying iceberg lettuce from California reached New York, the quality of the cargo left much to be desired.

Although some salad greens and even tomatoes were grown in greenhouses, supplies of hothouse-grown produce were limited and prices were costly. The supplies of fresh fruits were equally meager: A few tired old apples and pears that had been stored in October were available in late winter and early spring, as were a few bananas and citrus fruits. Occasionally you could find some grapes that had been shipped from Spain.

Today, in every area of the nation including Alaska, it is theoretically possible to enjoy almost any fresh fruit or vegetable if you are willing to pay the price. However, geographic location, density of population, regional and ethnic preferences, affluence, and especially demand all play a role in determining availability. Heavily populated, affluent areas located near major airports have access to all type of fresh produce in or out of season. Lightly populated areas that are not near large airports have less access to out-of-season items.

Regional and ethnic preference also come into play. Okra and collard greens may be big sellers in Dixie, but there is little or no demand for them in America's heartland. Tropical vegetables are staples in parts of the country that have large Latin-American populations and Oriental vegetables are always available in areas with large Asiatic populations. But tropical vegetables and Oriental produce are expensive specialty items, if sold at all, in most areas of North America.

Financial makeup, too, will determine whether or not a particular item is available in your area. In affluent neighborhoods, high-priced, exotic goods are in great demand. In those places where the main concern is to get enough food on the table, there is little or no call for luxury items. Keep in mind, though, that if you can't find the produce you want where you live, some enterprising entrepreneur will make the product available if there is enough demand to turn a profit.

HOW TO READ THE PRODUCE ADS

Food-industry writers seldom call a spade a spade. In the lexicon of the supermarket ads, a shovel is an earthmover. If you want a tin of great big black olives and buy one labeled large, you will be disappointed. In order of decreasing size, ripe olives are labeled special super, super colossal, colossal, super jumbo, jumbo, super mammoth, mammoth, and then large. Super giant and giant are also available, but I am not sure where they fit in the heirarchy.

Advertising and labeling language are equally misleading. Vague descriptions such as "Farm Fresh," "Home Grown," "Fresh Picked," and "Hand Selected" are meaningless. As a rule, such California produce as carrots, celery, and melons are superior to those grown in other areas. If an ad reads "Western," it means it was grown in Texas rather than California. Our best pineapples are grown in Hawaii. If an ad says "Tropical

Fresh" or "Plantation Picked," you can bet your boots they are not Hawaiian pines.

The most elastic and abused word in the produce ads is *large*. Larger than what? In the produce ads a large grapefruit is one that is bigger than an orange; a large orange is one that is bigger than a lemon. Some food-store chains have recently begun to use numerical fruit sizes in their ads. This is certainly a step in the right direction, but because of a lack of conformity, the numbers can be very confusing at first.

In order to understand the numbers, you need to know how the fruit is packed. Grapefruit and Florida oranges, for instance, are shipped in standardized cartons that hold four fifths of a bushel. A grapefruit labeled SIZE 36 means that it takes 36 grapefruit of the same size to fill the carton. An orange labeled SIZE 80 means that it takes 80 oranges to fill the carton. Common grapefruit sizes, from smallest to largest, are 48, 40, 36, 32, 27, 23, and 18. Orange sizes are 125, 100, 80, and 64. Limes, however, which are sized in a similar way, are packed in smaller cartons that hold only one fifth of a bushel, so a size 36 lime isn't nearly as large as a size 36 grapefruit.

Cherry and plum sizes are based on a different system. In the days when they were hand-packed, the number of pieces of fruit in a single row in the top layer was the indication of its size. Cherries, now shipped loose in 20-pound containers, are still labeled 12-, 11½-, 11-, 10½-, and 10-ROW. Plums, even when poured into 28-pound cartons rather than cell-packed, are labled 4 × 5, 4 × 4, and 3 × 4.

The general principle to remember is: the smaller the numeral, the larger the fruit. The only exception to this rule is for fruit measured in inches. Some types of fruit are sent to market in bulk rather than cell-packed and sized according to diameter. Clearly a 2½-inch peach is larger than one labeled 2 inches. But fruit measured by inches is the exception rather than rule.

WEIGHTS AND MEASURES

Although we now have efficient electronic scales that are sensitive enough to weigh a feather, the produce business still uses the archaic system of dry measures that was established in England more than five hundred years ago: 2 pints equal 1 quart; 8 quarts equal 1 peck; 4 pecks equal 1 bushel.

Fresh produce is delivered to wholesale markets in bushels, yet for any one item there can be a fairly wide range in its weight. A bushel of green beans can weigh from 24 to 30 pounds.

Pints can also vary in net weight. In some retail markets you can buy a pint of strawberries that only weighs 10 ounces, yet a pint of strawberries from another store will weigh as much as 15 ounces. You can expect comparable net-weight differences in any item sold by dry measure: cherry tomatoes, raspberries, blackberries, and, to a lesser degree, blueberries.

These differences in the net weight of the contents of a pint basket make it difficult if not impossible for a consumer to comparison shop these items. A store selling strawberries weighing 10 ounces for 79¢ per pint is pricing higher than a store selling 15-ounce net-weight pints for 98¢.

Why the diversity? Some blame can be placed on retailers who play it cute. Nearly all our strawberries are shipped from California and Florida in 12-pint flats that are close to 12 pounds in net weight (12 16-ounce baskets). Some retailers shake up the berries before repacking, gaining an additional three or four pints. The retailers say they go through the berries to remove those that are decayed, a claim that is occasionally valid. But more often than not, even when there is little or no waste, an unscrupulous retailer will still get 15 to 16 pints from a 12-pint flat.

Because most, if not all, states permit berries to be sold by dry measure and have no minimum net-weight requirements for a pint basket, this practice is not illegal. Until new weight standards are spelled out in statutes, check the weight of all produce sold in pints. You should receive at least 13 ounces net weight for strawberries and 15 ounces for blueberries and cherry tomatoes.

While consumers often get less than they bargained for when they purchase a pint of strawberries, they may get more than they bargained for when buying fresh salad or mixed greens. It's not unusual to pay a dollar or more per pound for tap water. The time-honored practice of adding water to greens sold by weight is almost universal. These watered-down items include curly endive, escarole, leaf lettuce, romaine, spinach, collards, mustard greens, and turnip greens. Retailers claim they wash and sprinkle the greens to remove dirt and freshen up the appearance of the product, but I have seen produce that was as fresh as a rose and as clean as a whistle being given a bath.

With rare exceptions, these greens as well as most fresh produce are dry when they reach the retail stores via the wholesale terminals or the chain warehouses. Prior to being put on display on the produce stands, moisture is added either by immersion in a tub full of water or by a liberal wetting down with a hose. Since greens can absorb water equal to 25 percent of their net weight, this practice also soaks the consumer.

SALAD BARS

Not too long ago self-serve salad bars made their debut in restaurants and were well received. Now the latest wrinkle in supermarket produce sections, their reception there has been spotty. They do best in areas that have a certain type of clientele: The salad bar is ideal for a one- or two-person household. With its advantage of allowing the consumer to purchase a little bit of this and a dab of that, a greater variety of salad items can easily be purchased by people living in small households.

Because most of the items offered are low in calories, salad bars complement new eating trends. They can be a blessing for senior citizens who like to shop for themselves and bring home a light meal and are a great attraction for office workers at lunchtime. While the food items offered at the salad bars sell at premium prices compared to similar items on the produce stand, many people feel the advantages are well worth the extra cost. However, if price is an object and you have to watch your pennies, stay away from the salad bar. If a salad bar is run efficiently, is well staffed, and offers fresh, top-quality products, it is an asset. If the supermarket uses it to unload shoddy produce, the salad bar is a liability.

REFRIGERATION

In Nature's master plan, minerals are finite but flora and fauna are programmed to self-destruct. All fruits and vegetables have built-in time clocks that determine the length of shelf life prior to decay. Some of these clocks tick at a far more rapid pace than others. Perishable items like raspberries start to break down immediately after harvest. Hardier items like winter squash endure for months with little change.

As fruits and vegetables break down, they gradually lose the desired flavor and texture. While this process is relentless, it can be retarded, but not halted, by proper refrigeration. Consumers who use their refrigerators wisely may limit shrinkage and decay to a minimum while retaining maximum flavor and texture.

As a general rule, with a few exceptions, fresh vegetables are at flavor peak the instant they are harvested and should be refrigerated as soon as possible. Fresh fruit is usually harvested while unripe, and has to be ripened for a few days at room temperature until it is ready to serve. Unripe fruit will not ripen properly under refrigeration, but ready-to-eat fruit must be refrigerated.

The average temperature in household and commercial refrigerators is about 35° F (give or take a couple of degrees). This is an ideal temperature for nearly all fresh fruits and vegetables, but is too cool for those few that are sensitive to cold.

All the tropical fruits (avocados, coconuts, mangoes, bananas, papayas, and pineapples) can't take sustained temperatures below 50° F. Neither can yams or tomatoes. There is no need to refrigerate dry onions, old potatoes (as opposed to new potatoes), citrus fruits, acorn squash, butternut squash, Hubbard squash, pumpkins, waxed rutabagas, and whole, uncut watermelons. These items do better if stored in a cool, well-ventilated area.

While immature tree fruits should be purchased while firm and then ripened at room temperature prior to refrigeration, apples should be refrigerated immediately after purchase. Never ripen an apple. Cherries, grapes, and berries should not be exposed to room temperatures. They do not ripen after harvest.

FROZEN FOOD

The prudent, knowledgeable food shopper uses the frozen-food section as a backup when shopping for fresh produce. There can be no question that any fresh fruit or vegetable in season is superior in flavor, texture, and probably nutritional value to its frozen counterpart. However, when an item is out of season, unavailable, overpriced, or below par in quality, frozen produce can be a welcome substitute.

The industry that most of us take for granted began only about fifty years ago. The concept of flash-freezing was fortuitously conceived in 1912 by Clarence Birdseye while catching fish in frigid Arctic waters. He sold the idea to General Foods, who launched the frozen-food industry in the 1930s. General Foods supplied a few grocery stores located in fairly affluent areas with primitive freezers and an assortment of frozen food. My family operated one of these test stores in Westchester County, New York. Unlike the thousands of frozen items now being offered, we had at most about two dozen. They included vegetables, fruits, uncooked fish, and some uncooked meat. Frozen juice, complete dinners, baked goods, fried potatoes, and pizzas were not yet available. The first frozen products caught on very slowly. We had just graduated from the old ice chest to the new electrical refrigerators, but home freezers or even refrigerators with frozen-food compartments had not yet arrived. Therefore, frozen food had to be purchased and used the same day. Most people also had difficulty accepting the premise that it was all right to drop the frozen product into boiling water without thawing it out first. It wasn't until World War II, when women entered the job market and, at the same time, fresh-produce prices soared, that the frozen-food industry really took off.

When using frozen food as a backup, be aware that while certain items are excellent, others don't quite cut the mustard. As a rule, fruits do not freeze as well as vegetables. For all members of the cabbage family, including broccoli, Brussels sprouts, cauliflower, and collards, the flavor and texture of the frozen product is almost as good as the fresh. Artichokes, asparagus, lima beans, winter squash, and baby whole okra also freeze well. Leaf spinach and peas are fair to middling. Green beans, wax beans, corn, and summer squash tend to break down and get soggy. All frozen berries get soft and soupy when thawed out. The trick to using them is to serve them while they are still partially frozen. Frozen rhubarb is quite good, especially when cooked with a package of frozen strawberries.

Even though they may carry a higher price tag, whenever possible buy whole frozen vegetables rather than ones that have been chopped, mashed, or cut. The frozen whole-leaf spinach is superior to chopped spinach and broccoli spears are better than chopped broccoli. Asparagus spears are nearly always tender, while the asparagus cuts are short on the tender tips and long on the woody, chewy cuts.

While the flavor of frozen orange juice concentrate can't hold a candle to that of freshly squeezed, it is often a worthwhile substitute, especially when fresh-orange prices go berserk. Fresh oranges are usually too expensive to squeeze in the summer months and not very good in quality in the early fall. The other pure-fruit concentrates—grape, grapefruit, apple, and pineapple—are fine products, but when buying any frozen concentrate (or fruit beverage that comes in containers or tins), read the label carefully. Make sure it spells out the word JUICE. Products with labels that read PUNCH, ADE, COCKTAIL, BLEND, NECTAR, DRINK, or BEVERAGE contain at best only 10 percent fresh fruit juice; the remaining 90 percent is sweetened tap water. If you read the small print on the list of contents of these concoctions, you will note that water is always the first ingredient listed, usually followed by sugar or corn sweetener. Food processors are required by law to list the ingredients in order of quantity. The old labels that used to read PORK & BEANS now read BEANS WITH PORK.

Even if the label contains that magic word JUICE, don't buy the product if sugar or another sweetener has been added. The processors claim that the sweetening agent is added to guarantee a more uniform, flavor-enhanced product, but more often than not sweeteners are added to mask the lack of natural sweetness of unripe fruit.

CONTROLLED ATMOSPHERE (CA) STORAGE

A discovery for extending the availability of perishables was made at the New York College of Agriculture at Cornell University in 1940. At first, controlled atmosphere storage was used primarily to extend the shelf life of apples, but in recent years pears and grapes have also been stored successfully.

Prior to CA, apples could be stored at best for only about four months, at which point they would get soft, break down, and start to decay. With CA, apples can stay firm and crisp for about ten months.

Our atmosphere is roughly 80 percent nitrogen and 20 percent oxygen. In CA, freshly harvested apples are sealed in refrigerated storage rooms and the supply of oxygen is decreased to about 5 percent. By monitoring the carbon dioxide in the air, the apples are anesthetized in a state of suspended vegetation. When the sealed rooms are opened from three to nine months later, the apples are as crisp and tasty as the day they were stored.

CA storage of fruit offers qualities not supplied by other methods of food preservation. The fruit loses none of its nutritional value and there is no change in texture or flavor. This magical process is performed without the use of heat, frost, salt, sugar, smoke, radiation, or questionable chemical additives.

WAX COATINGS

Mother Nature applies a thin dusting of a waxlike powder on some fruits and vegetables to protect them from direct sunlight and to help retain moisture. This dusting, which gradually disappears after harvesting, is called bloom and is a hallmark of freshness. Knowledgeable customers look for this bloom as a guide to quality when purchasing blueberries, grapes, plums, and some varieties of apples.

Growers also apply a petroleum-based synthetic wax coating to fruits and vegetables. Unlike Mother Nature's bloom, man-made wax has almost unlimited staying powers. This wax is applied not only for cosmetic reasons but to prolong the shelf life of a product. Without this protective wax shield to retain moisture, there could be substantial losses due to shrinkage and even spoilage, which in turn would lead to higher prices at the produce counter.

There is nothing new about the coating of produce. The Chinese were experimenting with such a process as far back as the twelfth century. In the United States the practice has continued with the blessing of the Food and Drug Administration for more than sixty years despite petitioning by health-oriented consumer groups to ban its use. The produce industry, claiming that the practice is vital to its interests, has applied pressure in the other direction. This squabble has been going on for years. Recently, the FDA has approved of an alternative compound, sucrose fatty acid ester, which is made from natural products—beef fat and sugar. This has opened up several new cans of worms. The new spray is made from tallow, a waxlike fat that comes from the hind quarter of beef. It is not kosher and therefore not acceptable to those who observe the strict dietary laws of Judaism. The beef content is also unsuitable for vegetarians and the sugar content poses a problem for diabetics. At the moment the new product is on hold and produce is coated with the petroleum-based wax.

All citrus fruits and most of apples, pears, avocados, melons, cucumbers, peppers, tomatoes, yams, squash, and rutabagas are either sprayed or dipped in wax. The quantity used ranges from a slight amount on the citrus fruits to a thick coating on the rutabagas. Even chocolate candies are sometimes sprayed with a similar wax. Particularly in the summer months wax may be blended with chocolate so that the candy can take the heat without melting. Not surprisingly, the end product often tastes like a candle.

Neither the man-made wax nor Mother Nature's bloom will rinse off in cold water, although the bloom will eventually evaporate. If you object to the man-made wax, you will have to pare the fruit or vegetable with a knife or vegetable peeler.

MARKETING ORDERS

In the United States we have, for the most part, a free, uncontrolled marketplace. The rise and fall of prices of fresh fruits and vegetables is controlled by the law of supply and demand. When this law is not interfered with, prices go up if the demand exceeds the supply, and they go down if the supply exceeds the demand. But not all fresh fruits and vegetables are sold in a free, uncontrolled marketplace. About fifty different items are subject to federal marketing orders and agreements that bypass the law of supply and demand.

Marketing orders first enacted during the Depression in the 1930s were necessary emergency measures taken to help prop up farm prices that had plunged to well below break-even levels for farmers and growers. With the prices of controlled items almost annually reaching new record highs, the marketing orders are now archaic.

The United States Department of Agriculture (which represents the grower and not the consumer) allows growers legally to restrict the free flow of fresh produce to market by a complicated system of prorating shipments. Under this system a grower who ships more than his allotment may be hauled into federal court and heavily fined. The marketing orders, which are used to establish size and quality standards and provide money for research and sales promotions, artificially prop the prices to the consumer. While the marketing orders may be a boon to the smaller growers, they are opposed by the larger growers, the retail food industry, and the consumers. Consumer groups label the orders unfair because when there is a bumper crop, and supply would exceed demand, the orders withhold huge quantities from the market and prices do not fall. However, if there is a short crop, there is no limit to how high prices may rise.

A worst possible scenario occurred in the marketing of the California navel orange crop in 1981. Thanks to most unusual weather, this crop of navel oranges was 40 percent larger in tonnage than any crop in history. Huge quantities of oranges were withheld from the market, most of which were either sold as cattle feed or allowed to rot. Despite the record size of the 1981 crop, the fruit sold at higher prices than those from the 1980 crop, which was about half its size.

THE TEN COMMANDMENTS FOR PRUDENT PRODUCE SHOPPERS

1. Always Look for Top Quality ❧

Ignore a low price tag if the quality of an item is below par. In the long run (and in the short run too), top quality, even though it carries a higher price tag, is a better buy than an item of poorer quality. At today's high prices, any waste at all is most costly.

2. Always Buy What Is at Peak of Season ❧

When fruits or vegetables are at season's peak, they invariably are of better quality (more flavorful and better in texture) and sell at affordable prices. For example, even though they are most costly in the summer months when they are out of season, the summer apples aren't nearly as crunchy and juicy as the ones that sell for a lot less during the fall.

3. Be a Selective, Rather Than a Haphazard, Shopper 🍃

Be a fussy shopper (in the trade, fussy shoppers are called cherry pickers). In an average supermarket display of fruits and vegetables, there is a wide range in the quality of the offerings. Whether the products are excellent, fair, or poor, they will carry the same price tag. Take an extra second or two to determine the quality and choose the best.

4. Be a Flexible Shopper 🍃

When you are being selective, you may not find the quality you're looking for, or the item you want may be overpriced. In either instance, shift gears and choose another fresh fruit or vegetable of better quality or one that has a better price. If you strike out, visit the frozen-food section.

5. Avoid Purchasing Fruit That Is Either Too Green (Immature) or Dead Ripe 🍃

Most fresh fruit will not ripen properly if it was picked when it was too hard and green. However, fully ripe, ready to eat fruit may have been bruised prior to your making the selection (this is especially true in self-service markets). There are exceptions to this rule: Select hard as a rock apples and ready to eat honeydews and berries. When shopping for fresh vegetables, especially greens, the fresher the product, the better.

6. Don't Pay Premium Prices for Oversized Fruit 🍃

As a rule, the less costly medium-sized fruit are a better buy than the higher priced larger fruit of comparable quality. The exceptions to this rule are cherries and blueberries. The larger Bing cherries and the larger cultivated blueberries will have better flavor and are well worth the extra cost.

7. When Shopping in a Self-service Market, Caress, Don't Press, the Fresh Fruits and Vegetables 🍃

Countless tomatoes, avocados, bananas, mangoes, papayas, peaches, pears, melons, plums, apricots, and nectarines are destroyed by well-meaning but overzealous shoppers. Rough handling causes waste and spoilage, losses that are passed on to the consumer.

8. Use Your Refrigerator Wisely ❧

Most, but not all, fruits and vegetables should be kept refrigerated until ready to use. Some notable exceptions are bananas, pineapples, tomatoes, and yams. However, if your refrigerator is too cool to permit firm, mature fruit to ripen properly, ripen the fruit at room temperature and then refrigerate it. A fresh fruit bowl displayed at room temperature may add to the decor of a room but will probably cause the fruit to start to break down and lose flavor and the desired texture. Except for bananas and pineapples, keep ripe, ready to eat, fresh fruits in the refrigerator and serve them while they are still chilled.

9. Shop Around for Your Source of Supply and Select the Market That Constantly Carries Fresh Produce of Top Quality ❧

If you live in an area that has several competitive markets, buy your fresh fruits and vegetables from the one that has the best produce department. Even though most of the chains have upgraded their produce departments, some markets are head and shoulders above others. Even in two markets owned and operated by the same management, there can be a wide disparity in the quality of the produce. It's a good idea to get to know the produce manager of the market where you shop. Most of them will go out of their way to be helpful.

10. Read and Take Advantage of the Weekly Produce Ads ❧

Those items that are featured in the ads are usually offered at, or near, the wholesale cost. These weekly specials are used to entice you into the store. Along with the low prices, you'll probably find that the products on sale are of good quality because the stores usually feature produce that is plentiful and at peak of season. If the quality passes muster, the items on sale will be excellent buys.

Fruits and Nuts

Deciduous Fruits

APPLES
·
APRICOTS
·
CHERRIES
·
NECTARINES
·
PEACHES
·
PEARS
·
PLUMS
·
QUINCE

T he deciduous fruit trees grow in the earth's Temperate Zones. They have a three-month dormant season during which they shed their leaves. They yield but one crop of fruit per year.

There are two types of deciduous fruits. Those that have a single seed are known as stone fruits, or drupes. These include apricots, cherries, nectarines, peaches, and plums. The other type is multiseeded and includes apples, pears, and quince.

APPLES

Botanically, apples are members of the rose family and are related to pears and quince. They are grown in the temperate latitudes of all continents. In North America, the northern half of the United States and the neighboring areas of Canada are apple country. The only southern states that produce fair-sized apple crops are Virginia and North Carolina, and even here apples are grown only in the cooler, higher altitudes.

Apples are now available twelve months of the year. However, as late as the early 1900s, apples were not in season during the summer months. They were harvested in the fall and stored in cool areas. By late spring they were long in the tooth and would start to get soft and eventually decay. As better methods of refrigeration came into use, the apples were put in cold storage and had a longer shelf life. At the end of World War I, we started to bring in some from the Southern Hemisphere, where the seasons are the reverse of ours. The big breakthrough in extending the apple season occurred in the 1940s at the Cornell University Agricultural

College, when experiments with controlled atmosphere storage proved successful. Thanks to the CA process, fine, crisp apples are now available year-round.

While there are hundreds of documented varieties of apples, only about a dozen are grown commercially on a large scale. The top four varieties in the United States are the Red Delicious, Golden Delicious, McIntosh, and Rome Beauty, all of which are old standbys. (The Granny Smith, a new comer, is rapidly gaining popularity.) The Red Delicious accounts for more than 40 percent of the total tonnage of apples produced in the United States.

Red Delicious 🍎

This variety, which is America's favorite, is grown in most of the apple-producing areas of the nation. However, the best and biggest crops are grown in the Pacific Northwest. Washington State easily ranks as the top apple producer, and most of the apples grown in that state are Red Delicious.

Although the Red Delicious is our best-selling apple, it has some shortcomings. As far as appearance (looks, color, and shape), it has no peer. Though while it is long on looks, it falls a bit short on flavor and texture. Many less-popular varieties are tastier. The Red Delicious is only fair on flavor when eaten raw and is a very poor cooking or baking apple. Early in the season, the Red Delicious tends to eat woody. Late in the season it tends to eat mushy. Many consumers who insist that an apple isn't an apple unless it has a red skin are missing out on our excellent yellow- and green-skinned varieties.

Golden Delicious 🍎

This attractive and flavorful variety is also produced in most of the apple-growing areas, but the best are produced in the Northwest. Although it bears the same name and is similar in shape to the Red Delicious, it is an entirely separate variety. It is far juicier and sweeter and has a smoother flesh texture than the Red Delicious. It has a more tender skin and, unlike the Red Delicious, is a fine apple for cooking and baking. Have someone peel you a Red Delicious and a Golden Delicious and make a taste test. Nine times out of ten the Golden Delicious will walk away with the honors.

McIntosh &

The McIntosh is a very juicy, tender-fleshed variety. It is grown in most apple-growing areas. Unlike the Red and Golden Delicious, which are at their best when grown in the Northwest, the best Macs are produced in the Midwest and Northeast. The Mac is fine for eating out of hand, but it is not one of our better cooking or baking apples. Applesauce made from Macs almost has the texture of baby food. They cook up too soft to use in pies because the slices dissolve, and when baked whole in the oven they melt and collapse.

Rome Beauty &

This large, round, red-skinned variety is our top-selling baking apple. Despite its nice size and firm texture, it lacks flavor when eaten out of hand. However, the addition of sugar and spices sparks up the bland flavor when the Rome is baked whole in the oven. One reason the Rome is so popular for this use is that it holds its shape when baked. The trick to having perfectly baked Romes is to allow them at least an additional ten minutes in the oven after you believe they are done. The Rome can take a lot of heat.

Granny Smith &

This is a green-skinned variety that originated in Australia. It is the number one variety grown in the Southern Hemisphere; Australia, New Zealand, South Africa, Chile, and Argentina grow and export huge amounts of them. In recent years they have also been planted in the Northern Hemisphere, especially in the United States and France.

Not too long ago Granny Smiths were available only in the spring and summer months. They were brought in via boat to take up the slack until our native apples came back in season. When they first became available here, Americans resisted Grannies because of their green skin color. They were mistaken for Rhode Island Greenings, which are similar in color and used only as cooking apples. After a slow start, the Grannies, with their tart-sweet flavor and very crunchy texture, caught on and sold at premium prices. American growers, observing the success of the imports, got into the act. We now produce Granny Smiths in many areas, however the ones grown on the West Coast are superior to those grown in the East. CA storage combined with dual hemisphere production give us fine Granny Smiths twelve months of the year. Each year the Granny Smiths capture a larger slice of the market and, in the opinion of many in the produce industry, will eventually surpass the Red Delicious as America's favorite and best-selling apple.

Other Varieties 🍎

Some apple varieties that are not among the top five but are regional favorites include the Empires, Cortlands, and Jonathans. The Mutsu, the newest kid on the block, shows great promise.

The Empire is a new variety and was bred in the Cornell Agricultural College. It is a cross between a Red Delicious and a Mac and has inherited the better traits of both parents: It is as firm and crisp as a Delicious and as juicy as a McIntosh.

The Cortland is a fine-textured, good-flavored red apple similar to the Mac. It is primarily grown in the Northeast and is at its best early in the season, when it is firm and crisp. In mid and late season, the Cortlands tend to be on the soft side. They do well in CA storage, but not too many are stored in that manner.

The Jonathan used to be a big seller nationally, but today, except for in the Midwest where it is a regional favorite, it doesn't sell too well. It is a fine-flavored, fairly small red apple that has a rather tough skin and tends to break down in storage.

The Mutsu is a golden-skinned import from Japan that is now being grown in the Northeast. Except for its name, which leaves something to be desired (it's pronounced mutt-soo), this apple is close to being perfect. It has a fine flavor, a crisp, juicy texture, and it reaches tremendous sizes. I've seen some that were as large as a grapefruit. This is the newest variety on the market and it will take some time before it achieves national distribution and acceptance.

The Rhode Island Greening is our best pie apple and is available in limited quantity during the fall. Most of the greenings are purchased by canneries for use in applesauce.

A number of popular varieties of yesteryear have just about disappeared from the scene. Some were in demand only because of their longevity, not their flavor. Prior to cold storage, CA storage, and imports from the Southern Hemisphere, apples that could hold up into the summer months were in demand even if they lacked flavor. Others in this group just went out of style. Some, like the Baldwin and the Northern Spy, deserved a better fate and are sorely missed.

The Baldwin was the best all-around eating, cooking, and baking apple and can now be found only at a few roadside stands in apple country in New York State. The Northern Spy was another great all-purpose apple that has almost faded from the scene. They are available on a limited scale in New England, their area of regional preference.

Other varieties, like the Winesaps, Staymans, Macouns, Spitzenbergs, Winter Bananas (a most appropriate name for a golden-yellow apple), and York Imperials, have all but disappeared from the commercial scene although they may be available locally.

Unlike some of the other fresh fruits that hit flavor peak when they reach full ripeness, a ripe apple is worthless. A mature apple is soft, mealy, and dry. The ideal apple is one that is crisp, crunchy, and juicy. So select very firm, colorful, unbruised apples. The medium-sized fruit is usually the best buy. Even though larger apples often sell for more money, they are more prone to bruises and are less apt to be as firm as the medium-sized ones. However, undersized apples (less than 2¼ inches in diameter) are often too woody.

Unlike the rest of the deciduous fruits, apples are at flavor peak when just picked from the tree. They should always be kept under refrigeration and never (even in a fruit bowl) at room temperature. An apple that is allowed to get mellow loses the desired crunch and juicy flavor.

□ WHEN TO BUY: *At peak October through June*
□ WHAT TO LOOK FOR: *Very firm, colorful, unbruised fruit*
□ HOW TO STORE: *Refrigerate immediately*

APRICOTS

The apricot originated in China thousands of years ago. This drupe gradually worked its way westward via camel caravans, was transplanted and flourished throughout Asia. Botanically it is identified as *Prunus armeniaca* and used to be known as an Armenian apple or peach. Apricots were referred to in Greek mythology as the Golden Apples of Hesperides.

When the apricot was transplanted to California, it found an ideal home. California produces a huge annual crop, most of which is packed in tins or sold as dried fruit. Washington State also grows excellent apricots, and in recent years its crop has shown significant increases in tonnage.

The apricot, a freestone fruit, features a beautiful orange color, a velvety skin, a most delicate flavor and texture, and a lovely fragrance. Of all summer fruits they are by far the most fragile. At their flavor best when tree-ripened, apricots are like golden balls of sweet juice. A truly ripe apricot is almost like a liquid nectar in texture. However, when they are fully ripened they can't withstand the rigors of packaging, transportation, or marketing. So apricots picked to ship to other areas, even around the corner, are harvested while quite firm. If they are picked while firm but with nice color, they will ripen properly at room temperature and be quite flavorful. If the growers play it too safe and pick the fruit when it is quite green, rather than color up and ripen, the apricots will shrink and shrivel and have little or no flavor. Don't purchase fresh apricots that lack high color. Those that are light green may or may not color up. Those that are dark green will never ripen.

'Cots, as apricots are known to the produce trade, have an all too short season (almost as short as the cherry season). Those grown in California and Washington are at peak of season in May, June, and July. In December and January we import a few from the Southern Hemisphere and, considering the distance they have to travel, they arrive in surprisingly good condition.

Until a few years ago, only a few varieties of apricots were shipped to market. The old standbys included the Royals, Tiltons, Perfections, and Moorpacks. However, recently we have been deluged with many new, very colorful varieties. One of these, called the King, is almost as big as a peach. Another fine-looking new apricot is the Flaming Gold variety. It's still too early to pass judgment on these new kids on the block, but some of them look very promising.

☐ WHEN TO BUY: *At peak July and August*
☐ WHAT TO LOOK FOR: *Unbruised, high-color (orange) fruit*
☐ HOW TO STORE: *Ripen at room temperature; refrigerate when ready to eat (soft to the touch)*

CHERRIES

Cherries are drupes that are very closely related to plums and more distantly related to peaches and nectarines. They are tasty, colorful, nutritious, and require no preparation. If cherries have one flaw, it is that they are available for an all too short a period of time. The cherry season is short and sweet. Enjoy them during the months of June and July, because they probably won't be available any other time during the year.

Although cherries originated in the Middle East and have been cultivated for thousands of years in both Europe and the Orient, by far the biggest producer, user, and exporter of cherries is the United States. Most of, and the best of, our cherries are grown on the West Coast, with the state of Washington number one in production. The large crop from California starts off the season late in May and winds up in mid-June. Then the scene shifts to the northwestern states and winds up early in August in British Columbia. Sweet cherries are also grown in the midwestern and northeastern states, but their fruit doesn't compare in size or flavor with western cherries. (Michigan and New York State have large crops of sour cherries, nearly all of which are sold to canners. Only an insignificant amount is marketed as fresh fruit.)

When purchasing almost any other fresh fruit there are several good varieties to choose from, but when it comes to cherries, it's a one-horse race. The Bing variety is in a class by itself. The rest of the cherry varieties are either also-rans or never-rans. The Bing is the tastiest, firmest, meatiest, and largest cherry grown.

The Tartarians and the Burlatts are the two early varieties that precede the Bings to the marketplace. Although these first arrivals are usually quite costly, the quality and the flavor is mediocre and the fruit is flabby. It's a good idea to pass up the early cherries and wait for the arrival of the Bings, which are usually available the last week in May and rule the roost until nearly the end of July. As the Bing season starts to taper off, the later-blooming Vans, Larians, and Lamberts arrive. These three varieties are quite good but are not in the same league as the Bings either in size or flavor and especially in firmness.

There are two light-colored or white varieties of cherries—the

Napoleons (Royal Anne's) and the Raniers. These attractive cherries are cream-colored and sport a red cheek. They have good size and fairly good flavor. The white cherries are not as firm as the dark cherries, are more fragile and easily bruised, and have a short shelf life. They have never become too popular because just as some people will buy only red-skinned apples, most people expect cherries to be red.

When you shop for cherries, what you see is what you get. They will not ripen or improve in flavor after you make your purchase. If you buy pale, light-colored cherries, they won't be nearly as sweet as the darker-colored fruit. Always select the firmest, darkest, and largest cherries available. The condition of the stem is a clue to the freshness of the cherry. A fresh cherry will have a green stem firmly attached to the fruit. If the stems have started to discolor, and especially if they are no longer attached to the fruit, the cherries are showing their age. They were either delayed in transit or your retail market ordered them too far ahead. An aged cherry isn't nearly as firm as one that has been harvested more recently. Pass up any cherries that are soft and flabby. Especially avoid those that are sticky or wet. Any decay is contagious, and like the proverbial bad apple, the bad cherry will contaminate nearby sound ones.

With the exception of cultivated blueberries, in most other fresh fruit size has little bearing on flavor and texture. A medium-sized orange or apple is every bit as good as their larger counterparts. But when it comes to cherries, the bigger the better. Larger cherries are more costly than smaller ones, but have more flavor and better texture and are well worth the premium price.

So look for the biggest, darkest, firmest, and freshest Bing cherries available. They'll probably be costly, but it doesn't pay to play a waiting game. At best the Bings are in season for eight to ten weeks. If you miss the boat, except for an insignificant, but very costly, supply imported from the Southern Hemisphere in January, the next boat won't dock for another forty-two weeks.

☐ WHEN TO BUY: *At peak June and July*
☐ WHAT TO LOOK FOR: *Large, dark, firm fruit*
☐ HOW TO STORE: *Refrigerate immediately*

NECTARINES

The nectarine originated in ancient China. It is botanically classified as a drupe and is akin to the peach, plum, apricot, and almond. While there is some difference in opinion among pomologists as to whether the nectarine is a fuzzless peach, a cross between a peach and a plum, or a distinct variety, there can be no question that it is one of our finest flavored, most succulent summer fruits.

Ninety-five percent of our total output is produced in California's San Joaquin Valley. Their season lasts from May through September. In December, January, February, and March we import nectarines from Chile, where they grow the same varieties as we do in California and are getting better at it each year. Despite the long distance that the Chilean nectarines have to travel, some of this fruit compares favorably with our finest California nectarines. As a rule, these imports bring top dollar because they arrive when there is a dearth of other fresh fruit in the market. The only months of the year when nectarines are not available in the market are April, October, and November.

Nectarines were not always as popular as they are today. Until the late 1940s, nectarines that were commonly available were rather drab and pale green with a red cheek. They were small in size, white in flesh, and clingstone. They were very juicy and quite tasty with a pleasant tart-sweet flavor, but had one bad flaw: They were very fragile and tender. Because they bruised very easily and had a very short shelf life, nectarines were grown on a very limited scale and shipped locally, random-packed in half-bushel baskets via unrefrigerated vehicles. Often the fruit was damaged in transit. When they arrived at the stores, decay set in quickly because of the lack of proper refrigeration.

Today nectarines are a far cry from those of yesteryear. Their flesh is golden rather than white. They no longer are fragile, have a long shelf life, and are carefully handpicked, precooled, graded by size, packed in cushion-celled cartons, and shipped to market in refrigerated trucks.

The first golden-fleshed nectarine was developed in Le Grande, California, and made its debut in 1942. More than one hundred new varieties have been developed from the original Le Grand, which was a large, not too colorful clingstone, but its claim to fame was its golden-colored flesh. The original Le Grand and a later-blooming variety called the Late Le Grand are still shipped to market, but since the newer varieties have better color and flavor and are superior in yield per acre, the Le Grands are being phased out.

Domestically grown nectarines are now in season for five months of the year. The California season is kicked off early in May with the arrival of the Armking variety. While this variety isn't nearly as good as some of the later-blooming varieties that follow, the Armkings are in great demand and are usually quite costly as a reward for their early arrival. In the month of June, far better flavored, juicier nectarines sell for half the price of the Armkings, so it may pay to wait. The last arrival of California nectarines are also not quite up to the quality of the ones in the market in June, July, and August. However, in September there are some superb nectarines from Pennsylvania and Washington State that are as good as any produced in California.

The flavor peak of season for nectarines is June and July. The varieties that arrive then are freestone and are sweeter, juicier, and more flavorful than the clingstone varieties that come to market in August and September. Although a few undersized nectarines from Florida and Georgia precede the earliest arrivals from California, they are at best only fair in quality. South Carolina, which grows some of our finest peaches, is expected to come on line in a few years.

The different varieties are not identified at the retail outlets. However, for the record, the best varieties are freestone and arrive in June and July. In order of arrival they are Early Sun Grands, Firebrights, Independence, Sun Grand, Moon Grand, Flavor Top, and Fantasia. They are replaced in mid-August by the clingstone varieties which, although they have better size, aren't quite as tasty and juicy. These include Red Grand, Le Grand (the grandaddy of them all), Gold King, Late Le Grand, and September Grands.

Thanks to experimental work by growers and agricultural schools, we have countless new varieties of nectarines. Of all the fruits that have been tinkered with, by far the best results have been achieved with this one. The improved nectarines capture a larger share of the summer-fruit market every year. People in the produce industry predict that by the year

2000 the nectarine will have surpassed the peach as our number one stone fruit.

As superb as the nectarines that you purchase in the peak of season are, they don't quite compare with the flavor and succulence of those that are tree-ripened. However, a tree-ripened nectarine is too fragile and delicate to ship around the corner, let alone across the country.

When shopping for nectarines, select those that are highly colored, velvet-skinned, unbruised, and unblemished. Buy them while they are quite firm and allow them a day or two to ripen at room temperature. When they are ripe, and not before, store them in the refrigerator. Chances are that if you purchase ready to eat nectarines they may be bruised. Size doesn't play too important a role as a clue to quality, but the medium-sized fruit are usually the best buy. Do avoid very small fruit because they may have been picked before reaching maturity. Oversized fruit are more apt to be bruised and nearly always carry a higher price tag.

☐ WHEN TO BUY: *At peak (from United States): June, July, and August*
  At peak (from Chile): February and March
☐ WHAT TO LOOK FOR: *Colorful, unbruised, medium- to large-sized fruit*
☐ HOW TO STORE: *Ripen at room temperature; refrigerate when ready to eat*

PEACHES

Peaches date back to ancient China. They are drupes and are related to plums, nectarines, apricots, and almonds. They arrived in North America via Europe and the Middle East.

Our domestic peach season is from May to October. During the winter and early spring months, from December to April, a limited supply of peaches are brought in from Chile. Unlike other Chilean fruit, such as grapes which are very good, nectarines which are good, and plums which are fairly good, usually these imported peaches are not very tasty. Peaches are grown in most states, but California and the South are the largest growing areas. Peaches are also grown commercially on a large scale in New Jersey, Pennsylvania, Michigan, Indiana, and Washington State. For the most part, the peaches produced east of the Mississippi aren't quite as large as those grown in California, but the eastern fruit is usually more juicy.

The hundreds of varieties of peaches can be divided into two groups: the clingstones, also known as clings, and the freestones. In the freestone group are some varieties that aren't totally free and these are called semifrees. There is a great difference in the flavor, texture, and succulence of clingstone and freestone peaches. Ninety-nine percent of the cling peaches are grown in California and sold to canners who put up tinned peaches and fruit cocktail; nearly all the freestone peaches are sold as fresh fruit.

If you have ever bitten into a fresh California cling peach, you'll understand why they are sold to the canners. They are hard, rubbery, and not very juicy. But cooking them in sugar syrup changes the undesirable texture and makes them quite palatable. Some people object to the oillike texture of the sugar syrup in canned cling peaches, a thickness of the liquid caused by the high natural pectin content of cling peaches. You can avoid this by purchasing tinned freestones, which do not have nearly as much pectin. In Italy, cling peaches put up in wine are a great delicacy. In the fall, a small amount of California cling peaches are shipped to market and sold in predominantly Italian neighborhoods for similar preparations.

Peaches come in two flesh colors: yellow and white. Not too long ago, about half the peaches were white-fleshed, but, like the white nectarine, the white peach has been replaced by the newer golden-fleshed varieties. Back when we had lots of trolley cars but no jet planes, there were many fine varieties of white peaches. To my taste, the older, white varieties—Georgia Belles, Carolina Belles, Dixie Belles, and especially Hiley Belles—were the belles of the peach ball. What they lacked in color they more than made up with a distinctive tart-sweet flavor. They were much juicier than the yellow-fleshed varieties and had a more delicate texture. To many old-timers, the Belles may be gone, but they are not forgotten. A limited supply of a new white variety that is almost all red in skin is now being shipped to market. These command premium prices even though they are very fragile and easily bruised.

Today, as with nectarines and plums, there are countless new and improved varieties of peaches, and all of them are superior to the yellow fruit we had in the good old days. The varieties are too numerous to list and most of them are too close in appearance to call. My personal favorite is the Blake, but dozens of others are just as good.

When comparing peaches of yesteryear to those of today, the most noticeable difference is in the skin texture. Peaches of yore were as fuzzy as kiwi fruit. Today's peaches are as slick as a baby's bottom. Peach fuzz could not be rinsed off under the cold-water tap. It either had to be removed by peeling the fruit or it could be rubbed off with a damp cloth. People with sensitive skin would break out in a rash if they went near a peach. What happened to the fuzz? Do we now grow fuzzless varieties? No, today's peach is just as furry and fuzzy as the peach of yesteryear, but the fuzz is removed by mechanical brushing at the orchards prior to packaging and shipping.

The trick to selecting fine peaches is to look for firm, colorful, unbruised fruit. If they are too green in color (called ground color in the produce trade), they may shrink, shrivel up, and not ripen properly. Allow firm peaches a couple of days at room temperature to ripen properly. If you buy ready to eat peaches, especially in a self-service market, they may be bruised. Very small peaches should be avoided because they may have been picked prior to reaching maturity. The medium-sized peaches are every bit as flavorful as the larger and usually more costly fruit. The first peaches to arrive in spring are usually of below-par quality. They are undersized and overpriced. It is usually a good idea to wait until the month of June to start buying peaches and to stop in mid-September. By that time the fruit is coming out of storage, prices are higher than in mid-season, and the quality has started to slip. However, from June 15 to September 15—enjoy.

☐ WHEN TO BUY: *At peak June, July, and August*
☐ WHAT TO LOOK FOR: *Firm, colorful, unbruised fruit*
☐ HOW TO STORE: *Ripen at room temperature; refrigerate when ready to eat*

PEARS

The pear, a member of the rose family, like the apple and the quince, is believed to have originated in the foothills of Northern India and Afghanistan. They don't thrive in areas where the summers are too warm or the winters too cold, but are now grown in all parts of the world that have climates similar to that of their mid-Asian origin.

Our West Coast has the ideal climate and altitude for growing pears. California produces more than 50 percent of our huge domestic tonnage. The combination of California, Washington, and Oregon accounts for more than 90 percent of our total crop. A fair amount of Bartlett and Bosc pears are produced in New York State and Michigan, but the yield from these areas can't match that of the West Coast in size, appearance, and flavor. Most eastern pears are sold locally and used for home canning. Until just prior to World War II, domestic pears were in season from mid-July to mid-May and out of season during the late spring and early summer. Today that slack is more than adequately taken up by imports from the Southern Hemisphere (Argentina, Chile, New Zealand, Australia, and South Africa), and fine pears are available twelve months of the year. Some varieties, which due to their longevity were popular prior to the availability of imports, have now been phased out.

Pears, like apples, must be harvested while they are firm, long before they reach full maturity. Unlike peaches or mangoes, which are at their flavor best when are allowed to tree-ripen, a tree-ripe pear will eat soft and mushy.

While countless varieties of pears exist, only about a handful are now grown commercially on a large scale. The top domestic varieties in order of volume, not necessarily of flavor, are the Bartletts, Anjous, Bosc, Comice, and Seckels. The imports from Chile, Australia, New Zealand, and South Africa are mainly of the Bosc variety and Packham Triumph variety, which resembles our Bartlett. The domestic Forelle pears have only a minimal share of the market. The Winter Nelis, Easter Buerres, and Keiffer pears are no longer grown in commercial quantity.

Bartlett 🍐

The Bartlett made its debut in England in 1770. It was then known as the Williams pear and in many parts of the world it is still known by this name. Some fifty years later, when it was introduced to North America, it was renamed the Bartlett. As either Bartlett or Williams, this bell-shaped variety is the world's best-loved, best-selling, best-flavored, most fragrant, best-looking pear.

The Bartlett, the first pear of the new season, arrives in the market late in July. Its season winds down in December. They are called summer pears, as opposed to the later blooming varieties that arrive in the fall.

Huge quantities of Bartletts are packed in tins by the canneries in California. This is also the only variety that is sold as a dried fruit.

When shopping for Bartletts, choose those that are light green in color rather than dark green. They are dark green when harvested. As they ripen they go from dark green to light green to pale yellow to golden yellow. When they are overripe, they turn brown. Some retail stores sell ripe, ready to eat pears, but purchasing ripe pears is asking for trouble. If the retail market is a self-service store, chances are that overeager shoppers will have pressed and bruised the pears prior to your making the purchase. If the ripe pear gets by the other shoppers without getting bruised, the omission will probably be corrected at the check-out counter. The best way to ensure unbruised pears is to buy those that are firm and allow them to ripen at home by leaving them at room temperature for two or three days.

The very first of the California Bartletts to arrive at the market are pears that have been grown in the warm lowlands. These are called River Bartletts and go from too firm to too soft in a very short span of time. They seem to melt overnight. Too ripe pears eat mushy, so always wait three or four weeks after the first shipment of Bartletts from California before purchasing them. By then the Bartletts being offered will be ones that have been grown in the cooler foothills at higher elevations. These are called Mountain Bartletts and will remain firm and juicy even after they reach full color and desired ripeness.

Occasionally a Bartlett pear will sport a red cheek. Although very pretty and perhaps more desirable for a fruit tray or basket, they aren't any tastier than those without the more attractive coloring.

In recent years growers have developed a new strain of Bartlett pear that is completely red, with no trace of green or yellow. This beautiful pear is highly prized for use in centerpieces and baskets and usually commands a premium price. Its beauty, however, is only skin-deep and the Red Bartlett isn't quite as juicy and flavorful as the traditional yellow or the yellow with a red cheek.

The Bartletts are in season from late July through late December.

Anjou (Buerre D'Anjou)

The Anjou made its debut in Belgium in the nineteenth century. This variety isn't quite as tall as the Bartlett and is more oval in shape. It has a clear, light green skin that shows a trace of yellow as it becomes fully ripe. The flesh is fairly smooth in texture but isn't quite as juicy and flavorful as the Bartlett or the Comice, and it is not as tasty and crunchy as the Bosc. The best thing going for the Anjou is its modest price. They usually sell for about one-third less than the other pear varieties. This moderate price tag reflects the huge volume of Anjous usually in the market. It is the most abundant of the winter pears. The Anjou is a good variety to bake whole in the oven, as you would bake an apple, since the delicate but bland flavor is enhanced by the addition of sugar and spice.

Anjous are in the market from October to May. In recent years some Anjous have been put in CA storage and are still in good shape as late as July.

Comice 🍐

Comice pears, which made their debut in France around 1850, aren't the prettiest variety, but beneath their drab exterior lies the most juicy and possibly the most flavorful pear in the world. This delicately flavored fruit is prized by gourmets and served in Europe's most posh *haute cuisine* restaurants as a dessert pear. The largest and the best Comice are skimmed off and sold in costly gift packages via mail order.

The Comice is squat and chubby and has a dull, usually mottled, light green skin. It occasionally sports a red cheek. Because it is a most difficult variety to grow without skin blemishes, a fairly large percentage of the annual crop doesn't meet USDA standards for US #1. However, the skin discolorations and blemishes do not mar the flavor.

The Comice arrive in late October and are usually sold out by Christmas.

Bosc (Buerre du Bosc) 🍐

The Bosc pears got their start in either France or Belgium (both claim the honor) in the nineteenth century. This pear, which is russet in color when ripe, is our most shapely variety. It has a symmetrical body that tapers to a long slender neck. It is quite sweet and has a nutty flavor and texture. Although it bears the word *buerre* in its official title, the Bosc isn't nearly as soft and buttery or as juicy as a Bartlett, Comice, or even a ripe Anjou. It is at its flavor best when used while it is still quite firm and is crunchy yet sweet.

This variety, more than others that have a similar problem but to a lesser extent, ripens from the inside out. If you wait until it yields to gentle thumb pressure it will probably be overripe, will have started to break down near the core, and will eat mushy.

Domestic Bosc pears usually arrive in October and are at their best until February, but they are still available until well into May. Some that have been kept in CA storage are still available into July. However, the Bosc imported from the Southern Hemisphere arrive in March. The knowledgeable shopper will switch from the domestic Bosc to the new crop imports as soon as they become available.

Seckel ❧

Seckels, an American product, are the smallest of the pear varieties. The first Seckel pear tree was discovered growing near the Delaware River in Pennsylvania around 1800. These pears are a dull olive-green in color and often sport a red cheek. While they originated in the East, those grown in that area of the country (mostly New York and Michigan) are much smaller and usually more woody than those grown on the West Coast. The western Seckel is one of our tastiest as well as one of our most expensive pears.

Western Seckels are in the market from September through January.

Forelle ❧

The Forelle is another miniature variety. While it is as pretty as a picture, this petite, shapely, freckled, red-cheeked pear is not very flavorful. At best it has a bland, sweet flavor and is quite firm even when ripe. At worst, and this happens quite often, it is very astringent. You can't tell by its appearance but must wait until you bite into a Forelle to know whether the fruit is sweet or puckery. Forelles are primarily used in fruit decorations, usually with lady apples, or to add color to fruit baskets.

Packham Triumph ❧

The Packham Triumph, an important variety in the Southern Hemisphere, is very similar to our Bartlett. It is sweet and juicy and has a green skin that turns yellow as it ripens, but it never has a red cheek. It has a similar shape but a more slender neck than the Bartlett.

Imported Packhams, along with Boscs, are in the market from March through July, at which time the new crop of California Bartletts takes up the slack.

Clapps Favorite 🍐

The Clapps Favorite is a small- to medium-sized pear that has the coloring but not quite the flavor of a Bartlett. They are primarily grown in the East and Midwest and are mostly used as a cooking and stewing pear. They are also used for home canning.

Winter Nelis 🍐

The Winter Nelis are small to medium, chubby, greenish-brown pears. The skin is often rough to the touch because of russeting. It is only fair in flavor. Prior to the imports, it used to be an important variety because of its longevity. Winter Nelis were available in late spring, when the major varieties were out of season. Today, very few are shipped to market.

Asian (Japanese Pears) 🍐

The Asian pear looks like a russet-skinned Anjou and eats like a very crisp, juicy apple. It seems to be a cross between an apple and a pear, but is an entirely separate breed. Asian pears, which are also called Japanese pears, are, not surprisingly, of Far Eastern origin and are highly prized in the Orient.

Although grown in California, much of the small crop is exported to the Far East. The balance is shipped to the wholesale terminals in the larger cities and purchased by merchants who supply stores in China-towns. Since the demand always exceeds the limited supply, this unusual but tasty fruit always sells at lofty prices, usually about twice the going rate for fine apples or pears.

Japanese pears, like apples, are best when very firm. Always store under refrigeration.

All varieties of pears should be purchased while still firm and allowed three or four days to ripen at room temperature. When they have almost reached full maturity (the Bartletts will have attained a pale yellow color; the Comice, Anjous, and Seckels will have a slight yield to gentle pressure; and the Bosc will have attained a milk-chocolate brown color but will still be very firm), put the pears in the refrigerator. Remember that a not quite ripe pear may be further ripened, but an overripe pear has passed the point of no return.

☐ WHEN TO BUY: *Available year-round*
☐ WHAT TO LOOK FOR: *Firm, unbruised fruit*
☐ HOW TO STORE: *Ripen at room temperature; refrigerate when ready to eat*

PLUMS

Plums are native to the Orient, Europe, and North America. While they date back hundreds of years, most of the ones we now enjoy are new improved varieties that have been introduced in the last four decades. There are countless documented varieties and new ones continually come on line.

Botanically the plum is a drupe and is related to the peach, nectarine, apricot, and almond. Plums come in assorted sizes, shapes, skin colors, and even flesh colors. Some varieties are slightly larger than a marble, others are much larger than a jumbo egg. They can be either round or oval in shape. There is a wide range of skin colors: green, yellow, orange, purple, every shade of red, and black. Most plums have a yellow-colored flesh, but a few varieties have a red flesh. Most are clingstone, but several are freestone.

Even with all this diversity of color, shape, and size, plums can be divided into two distinctively different groups that are known as European plums and Japanese plums.

The European varieties are usually, but not necessarily, freestone. They all have yellow flesh, but the thing that sets them apart is that they always have a purple skin (sometimes it looks almost blue if it has a lot of bloom).

The Japanese plums come in a wide range of colors but never have the telltale purple skin color. Most varieties are also yellow in flesh, but some have a bright red flesh. Unlike the European plums, which for the most part are freestone, most, but not all, of the Japanese varieties are clingstone.

The Japanese plums are usually tastier and juicier than the European varieties. They are nearly always eaten out of hand or used as a fresh fruit, since they dissolve when cooked and are not suitable for drying in the sun to be marketed as dried fruit. The European plums have a milder flavor and a meatier texture. Some varieties are quite good when eaten out of hand, but most are at their best when cooked or baked. These are the only types that are sun-dried and sold as dried fruit.

The terms *plum* and *prune* always cause some confusion, possibly because the Latin word for plum is *prunus*. Botanically there is no difference between a plum and a prune. Originally in the English language, the words *plum* and *prune* were used as synonyms. Today, to make it less confusing, the Japanese varieties are called plums and the European varieties are usually called fresh prunes. The sun-dried fruit are also called prunes.

Domestic plums are in season from mid-May until October. We import plums from Chile during January, February, and March. Almost 90 percent of our plums are grown in California. The first California plums of the year arrive near the end of May and are of a variety called the Red Beaut. This is not one of our better eating or holding varieties but is in great demand and brings top dollar because it is the first plum on the scene. You'll do better to skip these first arrivals and wait about three to four weeks for the more flavorful, better textured Santa Rosa plums.

Except for a jade-green–skinned variety called the Kelsey, the best-flavored Japanese-type plums are either bright red or dark red in skin color. The dark red varieties are round and have very small pits. They have good size, lots of juice, and a refreshing tart-sweet flavor. The top varieties in this group, in order of merit, are the LaRoda, Eldorado, Queen Anne, Friar, and Nubiana. The brighter-red–skinned varieties are usually more oval in shape. They have larger stones than the dark red varieties and a slightly more tart, but also delicious, flavor. The varieties in this group are easy to recognize by name because they usually end with the word *Rosa*. There are the Santa Rosa, Queen Rosa, Gar Rosa, and Simka Rosa. The Simka Rosa is the largest fruit in the Rosa group and is sometimes called the New Yorker Plum. (Over the years this variety has increased the income of the dental profession. Along with the fairly large and very obvious pit that is easily removed, there is usually a tiny chip of a pit embedded in the flesh of the plum. It lurks unnoticed until you bite into it.)

All of the above-mentioned varieties have a yellow flesh. There are some fine red-fleshed varieties that are as flavorful as they are colorful. The old Duarte variety, which was the original red-flesh plum, lost favor not because it lacked flavor but because its skin would split open in several areas as it reached full maturity. The newer red-fleshed varieties are very popular in California but are seldom shipped out of the state because they are quite fragile when ripe and are often too tender to risk lengthy transport. The red-fleshed varieties include the Frontier, Mariposa, Carol Harriss, and Ace, which looks like a huge Duarte. One red-fleshed variety warrants special attention because its name accurately describes its contents: It is called the Elephant Heart.

There are two green-skinned varieties that also warrant special attention. They look somewhat alike in color and are very much alike in shape and size. As they ripen, the green fades to yellow and then bronze. The lighter green variety, which arrives late in June, is called the Wickson and is about as dry and flavorful as an unsalted boiled potato. The jade-green variety, which follows in a few weeks, is called the Kelsey. There is no finer plum than the Kelsey. If you buy green-skinned plums, ask for Kelseys by name and avoid the Wicksons. Very often these two varieties are marketed as Green Gage plums, which they are not. The true Green Gage is grown in England and is a smaller horse of the same color.

The European-type plums don't reach the market until August. The most popular and the biggest crop are the small, purple, freestone prunes. These are grown not only in California but in the Pacific Northwest, Michigan, and New York. Washington State ranks number one in production. They are most commonly known as fresh prunes or Italian prunes but, depending on the ethnic balance of a neighborhood, may also be known as German prunes or Hungarian prunes. They are usually in good supply and sell at moderate prices. These freestone prunes are fairly good when eaten out of hand as a fresh fruit, but achieve their top flavor and top texture when used in baking. They also are excellent when stewed in sugar syrup.

One variety of European-type plum, called the Damson, is excellent for making jams and preserves. These are very small in size and are too firm and tart to eat raw, but they are fine when cooked. In Grandma's era, when many people put up homemade jellies, Damsons were a big item. Today they have all but disappeared.

In mid-September we get two very large-sized varieties of European-type plums. One of these, which resembles a huge fresh prune, is called the Empress and has a pretty good flavor. The other variety is called the President plum. These large purple plums are graceful fruits that look like winners but taste like losers. The Presidents arrive in huge supply just as the superior Japanese varieties are fading out. They are one of our poorest-flavored plums. When consumed while firm and green, they lack flavor and aren't very juicy. If allowed to ripen, they are very dry but do have flavor—they taste like soap.

Another purple-skinned variety to avoid is aptly named. This variety is called the Tragedy plum and they do have a tragic flavor. These small purple plums are grown in California and arrive in July about a month before the look-alike fresh prunes. Shoppers, mistaking them for the fresh Italian prunes, buy the Tragedies. They haven't nearly as much flavor raw or cooked. Thankfully, fewer Tragedies arrive in market each year.

When shopping for plums, select those that have good color, are at least medium-sized, and are unbruised. Remember to avoid buying the first plum varieties that arrive in market in early May. Wait until June for more flavorful, less costly varieties. Enjoy plums during June, July, August, and early September. When the red-skinned varieties go out of season in mid-September, don't substitute with the large, attractive purple President plums, which lack flavor.

Nearly all the stone fruits (nectarines, peaches, and plums) should be purchased when firm and at high color, then allowed to ripen for a few days at room temperature. Once they attain a slight yield (or give) to gentle pressure, they should be stored in the refrigerator until used. Remember that slightly underripe is preferable to overripe. The one exception to this rule is the apricot. You can't have an apricot that is too ripe. They are at flavor best when the texture is almost fluid.

- □ WHEN TO BUY: *At peak June, July, and August*
- □ WHAT TO LOOK FOR: *Firm, colorful, unbruised fruit*
- □ HOW TO STORE: *Ripen at room temperature; refrigerate when ready to eat*

QUINCE

The quince, which originated in Asia Minor, dates back to antiquity. A member of the rose family, it is related to both the pear and the apple and, like them, blossoms in the spring and bears fruit in the fall.

In the early 1900s quince was commercially grown in the north-eastern states, but today our major source of supply is California. While there are many varieties of quince, the type primarily grown in California is called the pineapple quince. This variety is shaped somewhat like a short-necked pear. When ripe the skin is yellow in color and smooth to the touch. However, when the quince is picked from the tree, the skin has a woolly, fuzzy texture. As with the peach, this fuzz is removed by mechanical brushing prior to shipment of the fruit to market. The quince is firm and most fragrant and will keep for months on end if stored in a cool area.

The quince is too firm, tart, and astringent to be palatable when eaten raw. However, when it is cooked in a sugar syrup along with a touch of clove and cinnamon, the firmness, tartness, and especially the astringency disappears. The light-colored flesh takes on a pinkish or amber hue when it is cooked and makes a delicious, fragrant, and colorful dessert. Quince jelly or marmalade, a taste treat in years past, has for some unexplainable reason gone out of vogue and all but disappeared from the shelves of the supermarket.

In the Victorian era, prior to chemical air fresheners, the fragrant quince was placed in clothes closets and bureaus to combat musty odors. Woolens, linens, and laces of that era smelled of lavender and quince.

☐ WHEN TO BUY: *At peak November and December*
☐ WHAT TO LOOK FOR: *Firm, unbruised, yellow-skinned fruit*
☐ HOW TO STORE: *No refrigeration required*

Citrus Fruits

GRAPEFRUIT

·

KUMQUATS

·

LEMONS

·

LIMES

·

BITTER ORANGES

·

MANDARIN ORANGES

·

SWEET ORANGES

·

UGLI FRUIT

C itrus fruits are produced on evergreen trees and are native to Ancient China and India. They thrive in subtropical climates, but don't do as well in the very warm tropics and are badly hurt in areas with even moderate frost. Citrus fruits are grown commercially in all the subtropical areas of the world. In the United States they are grown on a very large scale in Florida and California. Texas and Arizona also have substantial crops. Louisiana and parts of Georgia grow citrus on a very small scale. Although only six states out of our fifty grow citrus commercially, the total combined citrus fruits are our nation's number one fruit in tonnage and dollars and our number one export fruit crop. The five basic citrus fruits are: oranges, lemons, grapefruit, limes, and Mandarin oranges. Mandarins are hybrids created by agricultural husbandry. A few citrus varieties that don't quite fit into the basic five are: kumquats, ugli fruit, and bitter oranges.

Citrus fruits have a wide range in size, from the tiny kumquat to the huge three-to-four-pound pomelo. The pomelo is an ancestor of the modern grapefruit, but is not grown commercially. All citrus fruits are green in skin color prior to maturity and color up as they ripen. As a rule, ripe citrus fruits have either golden-yellow skins or orange-colored skins that are often mottled or russeted. By some quirk of nature, most ripe citrus fruits, if not harvested when they reach full color, revert back to green. An orange or grapefruit that is green because it has yet to reach full maturity and high color will be tart rather than sweet. But oranges or grapefruit that have begun to regreen late in the season will have a high sugar content. Citrus fruits do not ripen after harvest. When they age off the tree, they get softer but not sweeter.

GRAPEFRUIT

Today's superb grapefruit are a far cry from the original grapefruit, which were called Pomelos or Shaddocks. The Pomelos and Shaddocks had greater size, puffy, thick skins, lots of seeds, and very little juice, and were quite sour.

Grapefruit are grown in many areas of the world, with the United States, Israel, Spain, Greece, Brazil, and Cuba as the major producers. The United States is by far the world's number one grower of grapefruit, with Florida being our number one growing area, producing 75 percent of the United States crop. Texas ranks a distant second. California and Arizona also grow grapefruit, but on a limited scale. Though we export huge quantities to Europe and Japan, supplies for the domestic market are more than adequate.

Although grapefruit is one of Florida's most important cash crops, it is a fairly new enterprise. Grapefruit were first introduced in Florida in the early 1800s, but for one hundred years they were sold only to tourists as curios. It wasn't until the turn of the twentieth century that the first grapefruit were shipped, in limited supply, to the northern cities. Florida grapefruit are now shipped to all parts of the United States and Canada as well as Europe, the Near East, and the Orient.

Florida grapefruit are grown in two areas of the state. Those that are grown along the East Coast from just north of Palm Beach to just south of Daytona are called Indian River grapefruit. Those that are grown in central Florida are known as interior grapefruit. The Indian River section produces the finest fruit. Most of those grown in the interior are sold to commercial processors; they are rarely sold at retail markets. The few that are sold as fresh fruit are packaged in five- or ten-pound mesh bags and sold in supermarkets at modest prices.

Nearly all the Indian River grapefruit are marketed in retail stores or shipped overseas. Each grapefruit from this area is branded with an inked logo that reads INDIAN RIVER. The Indian River area, which runs parallel to the Gulf Stream, is perfect for growing citrus fruit. The warm ocean current provides ideal temperatures and shields the groves from frost damage when the mercury doesn't cooperate. Often when areas much farther south are hurt by severe frost, the Indian River section is spared because of its proximity to the Gulf Stream.

The best area in the Indian River section is an island called Orchid Island. There the world's finest grapefruit are grown. Every grapefruit shipped from this island bears a gummed label reading ORCHID or ORCHID ISLAND. Orchid grapefruit always top the market in price because they top the market in quality.

Texas also grows some excellent grapefruit, specializing in red-fleshed fruit known as Star Rubies or Ruby Reds. For reasons of geography and the resulting difference in freight costs, most Texas grapefruit are shipped to the Midwest and the West Coast. Very few are sold in cities along the eastern seaboard.

Grapefruit come in assorted sizes and in a variety of skin and flesh colors. Some have seeds, others are seedless. The skins may be golden-yellow, red-cheeked, bronze, or russet. The flesh colors are either yellow, pink, or red. Some grapefruit are as small as a fair-sized orange, others are as big as a melon—and they come in all sizes in between. Until about fifty years ago, nearly all grapefruit were of a variety called the Duncan. They were thin-skinned, heavy, fine-flavored, and full of juice, but also full of seeds.

Duncans are no longer shipped to market for table use but are grown in limited supply and sold to canneries and processors. These firms pack canned grapefruit sections and fresh or frozen-concentrated grapefruit juice. The Duncans have been replaced by a seedless variety called the Marsh Seedless. This variety originated from a chance seedling of the Duncan variety that produced seedless grapefruit. The Marsh seedless is a yellow-fleshed, seedless grapefruit that has fine flavor and texture and is fairly juicy, but it isn't quite as juicy as the Duncan. All the golden-fleshed grapefruit sold to consumers for table use are of the Marsh Seedless variety.

Mutations of the Marsh Seedless have yielded fruit that have a pink-rather than a yellow-colored flesh. These are known as Marsh Pinks. When darker pink strains were discovered and propagated, they yielded a red-fleshed grapefruit that we called Ruby Reds. In recent years in Texas, growers have come up with a new variety that is not only red-fleshed but has a reddish skin, called the Star Ruby. Other lesser pink-red varieties are known as Foster Pinks and Burgundy Reds. The pinks and the reds usually command a 15 to 20 percent higher price than the yellow-fleshed grapefruit. This premium exists only because the fruit is more colorful, not because it is juicier or more flavorful. Pink-, red-, or yellow-fleshed fruit of comparable quality are similar in flavor and texture.

Grapefruit that have clear yellow skins are known as Goldens. Those with slightly mottled skins are called Bronze, and those with heavily mottled skins are called Russets. As with the issue of flesh color, skin color plays no role in determining quality, and neither does size.

If neither size, skin color, nor flesh color are clues to quality, how does one judge a fine grapefruit? You have to check the weight and shape of the fruit, the firmness, the skin texture, the source of supply, and the time of year. As with all other citrus fruits, the thinner the skin and the heavier the fruit, the higher the juice content. A smooth as glass, slick-feeling, firm grapefruit is thin-skinned. One with a skin that has lots of pores, that doesn't feel smooth, or is soft or puffy has a thick skin. The shape of the fruit is also a clue to quality. A grapefruit that is flat at both the stem and blossom ends is ideal, but one that is round is acceptable. One that comes to a point at the stem end should usually be rejected since that portion of the grapefruit may not be very juicy.

Time of year also determines the quality, flavor, and texture of the fruit. While grapefruit are available twelve months of the year, the peak of season is from January to June. The Florida and Texas season starts in October. From October through December the grapefruit improves in quality, but it does not reach flavor peak until January. The Florida and Texas grapefruit season winds up in mid-June. In July, August, and September, the California-Arizona grapefruit come on line. At best they are only pretty good, and not nearly as good as those from Florida and Texas. During the summer months the price of grapefruit becomes quite costly.

☐ WHEN TO BUY: *At peak December through June*
☐ WHAT TO LOOK FOR: *Heavy, thin-skinned fruit*
☐ HOW TO STORE: *No refrigeration required if used within a week or two*

KUMQUATS

Kumquats are attractive miniature citrus fruits that are shaped like olives. They are highly prized in the Orient, where they have been cultivated for thousands of years. In the United States they are grown on a limited scale in Florida and California.

The fruit is quite tart and has many seeds. Its skin, like the skin of all citrus fruits, has a sharp, alcoholic flavor. Some people enjoy eating raw kumquats, skin and all, but most of those sold in North America are used for decorative purposes.

When used to make marmalade, the end product tastes very much like the English-style marmalades that are made with the bitter Seville orange. Kumquats are often glazed and candied with sugar and sold as a dried fruit. They are at their flavor best when boiled in a sugar syrup and used as a garnish for meat and poultry dishes, especially roast duckling. In Oriental neighborhoods, fresh kumquats, with their green leaves still attached, are a traditional New Year's gift.

When choosing kumquats, select those that are firm and have high color. If the fruit is orange in color, store in the refrigerator; if it is greenish in color, don't make the purchase. The peak of season is from December to May.

LEMONS

Lemons are available year-round. Most are produced in California and Arizona, but Florida has a small crop that is annually increasing in tonnage. Due to strict Federal Marketing Orders that restrict the free flow of California lemons to market, we sometimes import lemons from Spain, Italy, and Chile.

While there are several varieties of lemons, and they sometimes come from different areas, they are not identified by variety or source at the retail markets. Most of the California lemons are packed and marketed through a co-op—Sunkist Growers, Inc. Their top-quality lemons are individually stamped with the Sunkist logo. Lemons that don't quite meet their top standards are shipped to market unbranded. However, the top quality of the independent growers who are not members of the co-op also arrive to market unbranded.

The quality of lemons is judged by the color, clearness, and texture of the skin, not by the size of the fruit. As lemons age, the light yellow color turns to a darker yellow. Some lemons don't pass as #1's if they have skin blemishes or scars. The color and the clearness of the skin is no clue to its juice content even though those with blemished skins sell at lower prices at the wholesale (usually not the retail) level. However, the skin texture is all-important. The thinner the skin of the lemon (as well as any other citrus fruit), the higher the juice content. The smaller- and medium-sized lemons are usually thinner-skinned than the larger-sized fruit. The larger lemons are always more costly by weight than the smaller ones. As a rule, the less costly smaller lemons are a better buy. Gently rolling a hard lemon on a table will result in a greater juice yield.

Select firm lemons that feel heavy for the size and thin-skinned. Reject those that are light in weight, thick-skinned, or are soft or spongy. Lemons will keep for many weeks if stored in the refrigerator.

Lemons are usually inexpensive in the winter and expensive in the summer. Limes are inexpensive in the summer and expensive in the winter. Since lemons and limes are very similar in flavor and texture, the wise shopper will substitute with whichever is less costly.

LIMES

Limes are available twelve months of the year. The peak of season, which combines the best quality and the lowest price, is in July and August. Most of our limes are produced in Florida, a few are grown in California, and we also import some from Mexico and Venezuela. Most limes are either of the Tahiti variety or the Mexican variety, but most retailers identify their limes as Persian. However, I have yet to see a lime from Persia.

Limes are similar in flavor and texture to lemons. While they are not quite as tart, they are more fragrant than their yellow cousins. Since they are so very similar, in almost any recipe that calls for lemons, limes can be substituted, and vice versa. This is important for the produce shopper to know because sometimes there is a big difference in the prices of lemons and limes. As a rule, limes are dirt cheap in the summer months, when lemons are very costly—so don't hesitate to use limes in place of them. In the winter months it is wiser to use the less costly lemons in place of the higher-priced limes.

Limes are at their best when used to flavor beverages. They are also excellent for use on seafood, salad greens, and avocados. The fresher the lime, the darker green the skin color. A yellowish lime isn't too fresh and will lack acidity. As a lime ages even further, the skin will show brown, scalded areas. A yellow or even a scalded lime can still be used as long as it is firm, but it won't be as good as a dark green lime and should sell at a much lower price. Store in the refrigerator, but after a couple of weeks limes show a refrigerator burn on the skin.

BITTER ORANGES

Bitter oranges are also known as Seville oranges. Though they are not newcomers on the scene, they are seldom grown or used in the United States.

The flesh and the juice of the bitter orange are too bitter to enjoy as a fresh fruit. It is, however, highly prized for its use in making marmalade. Imported English marmalade is made almost exclusively from Seville oranges, but nearly all domestic marmalades are made from sweet California oranges. The difference in flavor is noticeable.

In the United States, a small crop of bitter oranges is produced in Florida and a few are also imported from Spain. On rare occasions they are offered for sale in specialty fruit stores. They look just like sweet Florida oranges and should be stored in the refrigerator. Unless you plan to make authentic English marmalade, it is no great loss if your favorite retail market doesn't handle bitter oranges.

MANDARIN ORANGES

The sweet orange and the grapefruit are the most important citrus-fruit crops. In third place is a large group of orange-skinned varieties of citrus fruits that are collectively known as Mandarins. There are more than a dozen varieties of them that are grown commercially. Those tiny Mandarin orange sections that are sold in tins at the supermarkets come from the Orient. These were first propagated as far back as 2000 B.C. Some of the newer varieties made their debut in Florida in the twentieth century. Most Mandarins (with rare exception) are flat in shape, have loose "zipper" skins, and peel and segment quite easily. Some varieties have seeds, some are seedless. While Mandarins are grown worldwide, nearly all those sold in North America are grown in either Florida or California.

Tangerines 🍃

Our best-known Mandarin is the tangerine, and most are grown in Florida. Early in the season, before the Florida fruit comes on line, we import some tangerines from Mexico. As a rule, the quality of this import leaves much to be desired. The most common varieties of Florida tangerines are the Robinson and the Dancy. Both are similar in texture, flavor, and color, although the Robinsons attain a larger size. Tangerines are not identified by variety at the supermarket fruit stands.

Tangerines arrive in October and are out of season in February. The early arrivals are usually a bit tart and those at the tail end of the season are often quite dry. Florida tangerines are at their best in December.

Murcotts (Honey Tangerine) 🍃

The Murcott is a much improved tangerinelike citrus fruit. It has recently been renamed and is now called the Honey Tangerine. These haven't been in the market nearly as long as the regular tangerines. They are firmer and juicier than tangerines but are more difficult to peel and segment and have a fair amount of seeds. The flesh is very sweet and has a rich, deep orange color.

Temple Oranges 🍃

The Temple orange is a cross between a tangerine and a sweet orange. They are shaped like, but are larger than, tangerines. Their loose-fitting skins make them easy to peel and segment, but they have lots of seeds. Temples have a fine, sweet flavor. They are not only excellent when eaten out of hand but render a very sweet, rich orange juice when squeezed. Florida Temples arrive in December and wind up in April.

Tangelos 🍃

The Tangelo is fairly new on the block. It is a hybrid resulting from crossing tangerines and other citrus fruit. They are firmer and larger than tangerines. They have tight, rather than loose, skins and aren't very easily peeled and segmented. Of the four varieties of tangelos grown in Florida, three have seeds and are shaped like a tangerine. The fourth is seedless and shaped like an orange. Of the four varieties, one is a loser, two are fairly good, and one is out of this world.

The first variety to arrive reaches the market in October and is known as the Early K. They are shaped like tangerines, have seeds, and are fairly juicy, but they taste like iodine. Fortunately, they have a rather short five-to-six-week season.

The next two varieties arrive in November and are called Novas and Orlando Tangelos. They are almost identical in appearance with the Early K. They have the same coloring, shape, juice content, and amount of seeds. However, unlike the Early K, both the Nova and the Orlando have a fairly sweet but slightly bland flavor. Both varieties arrive in market three to four weeks after the first arrival of the Early K.

The last tangelo to arrive, usually in mid-December, is called the Mineola. They are also called Red Tangelos, and in Florida they are called Honey-belles. The Mineola is a cross between a tangerine and a grapefruit. It is seedless, chock full of juice and flavor, and, in my opinion, the finest citrus fruit around.

Mineolas are easily identified. They are larger and more colorful than the other kinds of tangelos. Instead of being flat at both ends, they are round with a very identifiable nipplelike bump at the stem end. This is not only a superb table orange, but, when squeezed, it renders a very colorful, tart-sweet flavored juice. I am told that Florida growers personally use Mineolas to squeeze for orange juice.

California Mandarins ❧

In addition to Orlando and Mineola tangelos, California also grows several varieties of Mandarins. These include Kinnows, Fairchilds, and Satsumas. All three look like small, tight-skinned tangerines. As a rule, the California Mandarins have a higher skin color than those grown in Florida. (This deeper skin color is also true of California oranges.) However, most California Mandarins have a slightly tart flavor. I am told that this tartness is due to the practice of grafting Mandarin buds on lemon-tree stock.

Clementines ❧

Clementines are tiny Mandarins that are imported from Spain and North Africa. They are rapidly gaining favor in North America and an increasing tonnage is imported each year.

Though the Clementine is the smallest Mandarin, what it lacks in size is made up for with fine flavor and texture. Clementines are usually free from seeds and easy to peel and segment. Their tiny size, lack of seeds, and easiness to handle make them especially popular with very junior citizens.

The Clementines arrive in November and wind up in January.

☐ WHEN TO BUY: *At peak December, January, and February*
☐ WHAT TO LOOK FOR: *Firm, colorful, thin-skinned fruit*
☐ HOW TO STORE: *No refrigeration required if used within a week or two*

SWEET ORANGES

The United States is blessed with the world's finest oranges. They date back to when Florida and California belonged to Spain and orange trees were planted alongside numerous Spanish missions.

Since we grow oranges on both coasts, our fruit is available twelve months of the year. Once in a while we import Jaffa, or Shamouti, oranges from Israel; these are similar to our California Valencias. On rarer occasions we import oranges from Spain and North Africa. Imported oranges are but a drop in the bucket, with native supplies more than adequate to satisfy our own demand and maintain us as a major exporter.

Due to differences in soil and climate, there are differences in the color, texture, and juice content of California and Florida oranges. Even when both areas grow an identical variety, the end product is not the same. Florida oranges are thinner-skinned and have more juice than the ones grown in California, but they are more difficult to peel and segment. As a rule, the Florida orange is better for squeezing and the California orange is better for table use. Much of the Florida crop is used to make frozen concentrate or sold as fresh orange juice in paperboard containers and bottles.

The California oranges aren't quite as juicy and are not nearly as thin-skinned as the Florida oranges. Both the skin and the flesh of the California oranges are a deeper orange in color. All the California oranges and most of the Florida oranges are seedless, but one important Florida variety, the Pineapple orange, has lots of seeds.

Both California and Florida have two major crops of oranges each year. The early crop from Florida has two varieties—the Hamlins and the Pineapple oranges. The late Florida crop also has two varieties—the Valencias and the Pope Summers. The early California crop is of the Valencia variety and the late crop is known as the navel orange. Areas of Arizona that border California produce some oranges, and Texas, a major producer of grapefruit, grows a limited amount of oranges.

The first Florida orange to come to market arrives in October and is called the Hamlin. This is our only variety that doesn't bat a thousand. The Hamlins look like superb oranges—they are thin-skinned, seedless, and heavy in weight, and they have good color. However, they are full of pulp. They are pretty good when eaten as a table fruit, but they aren't very juicy when squeezed. What juice they do yield is very pale and watery and lacks flavor.

About four weeks after the arrival of the Hamlins, the Florida Pineapple oranges come to market. They aren't quite as thin-skinned as the Hamlins, but are just bursting with juice and flavor. This is one of our finest juice oranges. If they have one fault, it is that they have seeds galore. This variety is not grown in California.

During most of the month of November the look-alike Hamlins and Pineapples are both in the market. The Hamlin is a loser and the Pineapple is a winner. How can you tell them apart? You can't unless you cut into the orange. If it is seedless, it's a Hamlin and won't be very juicy. If it has seeds, it is a Pineapple and will have lots of juice.

The Pineapple oranges stay in season until mid-February. Late in January the first shipments of the Florida Valencias arrive. While the Florida Valencia is a superb juice orange, these first arrivals aren't quite mature and therefore not as sweet as the Pineapple oranges already in the market, nor are they as sweet as the Valencias will be later on in the season. The trick is to buy the Pineapple oranges as long as they are available. How can you tell these two look-alike varieties apart? Again, you have to cut open the orange. If it has seeds, it's a Pineapple; if it is seedless, it's a Valencia. However, even the earliest-arriving Valencia is a better orange than the Hamlin. The Florida Valencia at full maturity is our finest juice orange.

The Florida Valencias are joined in July by a later-blooming cousin called the Pope Summer. These Florida oranges are available until mid-August. In September and October the only orange in market for either juice or table use is the California Valencia.

The California Valencias arrive late in May and are in the market until December. The California navel oranges make their annual debut in November and are in season until June. Note that the California seasons, like the Florida seasons, overlap. The trick to getting the sweetest and best-textured California oranges is to buy the variety that has been in the market for several months, not the one that is just coming into season. For example, in May and June the California navels are superior to the newly arriving Valencias. However, in November it is the Valencias that are sweeter and juicier; the navels are not quite ready at that time.

The same pattern should be followed when purchasing Florida oranges: Always purchase the ones that have been on the market for the longest period of time.

Early in November and most of December, along with the California navel oranges, there is a limited supply of Florida navels in market. Just as the California grapefruit aren't nearly as good as those grown in Florida, the Florida navels aren't nearly as good as the ones produced in California. Florida navels have a light skin color and a light-colored flesh. They are thin-skinned and difficult to peel and segment. The flavor and texture isn't as good as that of a California navel.

Having covered the twelve-month cycle and the wisdom of purchasing the more mature variety when two crops overlap, here are some tips on how to identify top quality:

When shopping for oranges or any other citrus fruit, the hallmarks of quality are the thinness of the skin and the weight of the fruit in relation to its size. The color and the size are not clues to quality. The thinnest-skinned fruit will be heavy in weight and have more juice. Since the smaller- and medium-sized fruit are usually thinner-skinned and lower in cost, don't purchase oversized fruit.

☐ WHEN TO BUY: *Available year-round*
☐ WHAT TO LOOK FOR: *Firm, colorful, thin-skinned fruit that is heavy in relation to size*
☐ HOW TO STORE: *No refrigeration required if used within a week or two*

UGLI FRUIT

The name *ugli* is most appropriate, for this is the meanest-looking variety of the citrus fruits. It looks like a very poor grapefruit with a rough, loose-fitting, thick skin. The dull skin color is a mixture of green and russet-yellow. When fully ripe, the yellow portion of the skin takes on an orange-colored tint. The fruit feels spongy and the fruit comes to a point at the stem end. However, the unattractive, hidelike peel hides a surprisingly flavorful, juicy, sweet, fine-textured, pink-tinted flesh.

Ugli fruit is native to the island of Jamaica in the Caribbean Sea. It is probably an offshoot of the combination of a pomelo (the original grapefruit) and a sweet or Mandarin orange. The Jamaican crop is not very big and, since the supply is very limited, prices are always on the high side. Attempts to grow ugli fruit in Florida have not met with much success.

It is a fine-flavored fruit, possibly a shade sweeter than a grapefruit, and should be stored in the refrigerator. The question is, Are they worth two or three times as much as the cost of a comparable-sized grapefruit?

Tropical and Subtropical Fruits

❦

BANANAS AND PLANTAINS

·

CARAMBOLAS

·

CHERIMOYAS

·

FRESH DATES

·

FRESH FIGS

·

GUAVAS

·

KIWIS

·

LOQUATS

LYCHEES

·

MANGOES

·

OLIVES

·

PAPAYAS

·

PASSION FRUIT

·

PERSIMMONS

·

PINEAPPLES

·

POMEGRANATES

·

PRICKLY PEARS

T ropical fruits grow in the earth's warmer climes, in areas that are free from frost. These trees do not have a dormant season, nor do they shed their leaves, like the deciduous fruits. Most tropical fruit trees bear more than one crop each year.

BANANAS AND PLANTAINS

The wild banana, believed to have originated in Southeast Asia, is now cultivated in all the tropical areas of the world that have ideal conditions: warmth (never below 50° F), high humidity, ample and steady rainfall, and loose, well-drained soil. The areas that best meet these requirements are found in Central America. The top banana-producing countries of the world are Costa Rica, Honduras, Ecuador, and Panama. Except for a very small crop grown in Hawaii, bananas are not grown commercially in the United States.

The banana is a most unusual fruit, akin to no other. It is actually the berry of a giant, treelike herb. It grows on stalks that bear about seven to twelve hands of bananas. Each hand is made up of about twelve to fourteen bananas. A bunch, consisting of seven to twelve hands and up to a couple of hundred bananas, can weigh from fifty to one hundred pounds.

Unless you were born before the mid-1950s, you have probably never seen a bunch of bananas, unless you have visited the growing areas. Prior to 1964, bunches of bananas were a common sight in all retail markets. They were hung from the ceiling and the bananas were usually cut to order from the stalk at the time of purchase. These bunches were carried, one at a time, on the backs of laborers, into the holds of the banana boats in Latin America. When they reached port in the United States, they were unloaded in the same tedious fashion, one bunch at a time, on the backs of laborers, down the gangplanks and onto trucks waiting on the piers. A hundred years ago the same methods of loading and unloading were used, but instead of steam ships and trucks, the bunches were loaded onto sailing vessels and horse-drawn wagons.

In 1964 this crude, back-breaking, time-consuming method was scuttled. Instead, the hands of bananas were cut from the stalks right after harvest and packed in casket-shaped forty-pound boxes. These boxes were trucked to the piers, stacked on pallets, and lowered into the holds of the ships by mechanical cranes. Upon arrival at their destination, the ships were unloaded in similar fashion and the boxes of fruit were loaded onto trucks. Loading and unloading the banana boats under the old method used to take several days at each end. Today the ships are loaded and unloaded in a matter of several hours.

Bananas are available twelve months of the year. The supply is both ample and steady regardless of month, season, or weather, and prices are very stable and seldom fluctuate. Other fresh fruits that are far less difficult to grow and are produced much closer to home, usually sell for more than twice the price of bananas. But thanks to oversupply and fierce competition by the growing nations, bananas are nearly always the least costly item at the fruit stand.

Bananas are harvested while they are grass-green. (A stalk-ripened banana won't taste nearly as good as the ones you can buy in your neighborhood markets. A banana allowed to ripen prior to harvest is very soft, oily, and fragile.) When they reach port in North America, they are still grass-green and as hard as a rock. They are ripened, without being removed from the banana boxes, in sealed, temperature-controlled areas called banana rooms. These rooms are equipped to supply the required warmth and humidity to ripen the fruit properly. The bananas are exposed to a harmless ethylene gas, similar to the gas naturally exuded by all bananas, that accelerates the ripening process.

When the bananas are exposed to the proper warmth, humidity, and gas, they gradually change in color. They go from grass-green to light green to light yellow to golden-yellow. The light green or light yellow are the ideal colors to look for when purchasing bananas at the produce counter. After you take them home, the bananas will continue to ripen. Once they are golden-yellow, they will fleck with brown spots that look like freckles. The freckles will continue to grow in size, and eventually all the yellow color will have disappeared, leaving a totally brown-skinned banana. At each stage of the color change, the inner flesh of the banana will get softer and sweeter as the fruit ripens.

The trick to getting the most value and the least loss when purchasing bananas is to choose those that are light green or light yellow. In the produce trade this coloring is called on the turn. If you buy golden-yellow fruit that is ready to eat, you may have trouble. Golden bananas are quite fragile and often can't stand the rigors of a self-service type of marketing. Chances are that prior to your arrival at the produce counter, other over-zealous shoppers will have squeezed or bruised the fruit while making their selection. That part of the banana that has not been handled gently or has been bruised will discolor and turn black and will have to be discarded.

While bananas don't like cold weather, neither do they like it very hot. Once the temperature goes into the nineties, they tend to get very soft and seem to melt. Buying grass-green bananas is a big mistake, yet there is a tendency on the part of the consumer to buy them dark green in the heat of the summer. Very often if the bananas are too green, they will not color up or ripen no matter how warm the weather.

In the winter months, very often the bananas are green because they have been chilled. Chilled bananas never turn yellow but instead turn a dull greenish gray. They stay very hard and eventually wind up in the garbage can. Never buy jade green bananas.

Nearly all bananas offered for sale in most areas are the everyday yellow-skinned fruit, though there are also some red-skinned bananas that occasionally show up in some of the more exclusive higher-priced fruit shops. These coral-skinned bananas are shorter and plumper than the yellow-skinned variety. They have a softer texture and are slightly sweeter than the yellow bananas. However, they aren't worth the premium price unless you want to use them for decorative purposes. There is also a miniature variety called the Lady Finger banana. They have no special value other than their oddity. They are seldom if ever offered commercially.

A market with a fair amount of Latin American trade will carry fruit that looks like unattractive, oversized bananas, called plantains. Sometimes called the potato of the tropics, they are staple food products in all of the world's lush, tropical areas. This giant cousin to the banana is far too starchy to eat raw, but is very flavorful and smooth when cooked in any manner you would cook an Irish potato: boiled, mashed, baked, or fried. Thin slices of fried plantain are as tasty as our potato chips but far more nutritious.

Select light green or light yellow bananas and buy only as many as you will use in two or three days. They are especially perishable in the summertime because they ripen so quickly in the heat. The industry now advises that ripe bananas may be refrigerated for a few days and that though the skins will turn black, the banana itself will be fine. I belong to the old school. There used to be a jingle, sung by a young lady named Chiquita Banana, that advised never to put bananas in the refrigerator.

☐ WHEN TO BUY: *Available year-round*
☐ WHAT TO LOOK FOR: *Clear-skinned, unbruised fruit that is light green or light yellow (jade green or golden yellow fruit should be avoided)*
☐ HOW TO STORE: *Never refrigerate bananas*

CARAMBOLAS (STAR FRUIT)

The carambola is a strange-looking exotic fruit that resembles no other. It is native to Malaysia and is grown in fair supply in Hawaii and on the islands and lands that border the Caribbean Sea. In the continental United States it is grown commercially on a very small scale in Florida and California. Our production is limited not only because of lack of demand but because the trees are extremely sensitive to cold weather. They are at peak of season from September to January.

Carambolas are also known as Star Fruit because they resemble the five-pointed starfish, and the most common of the twenty known varieties is called the Golden Star. The edible skin appears to have a thick, waxy coating. Depending on the variety and the ripeness of the fruit, the flesh is either pleasantly tart or unpleasantly sour and has a few tiny seeds. As a rule, the deeper the yellow coloring, the less tart the flavor. Carambolas are usually eaten raw or used to flavor fruit punches and sometimes used to make jams and jellies. Allow them to ripen if pale yellow and refrigerate when the color turns golden yellow.

CHERIMOYAS (CUSTARD APPLE)

The cherimoya is native to the Andes Mountains of Peru and Bolivia. It is now grown commercially on a small scale in California. However, its annual sales show a very gradual increase. It is at peak of season during December, January, and February. Limited supplies and high prices exclude it from all but the very posh retail outlets.

The cherimoya has an unusual appearance. It looks somewhat like a barbless artichoke or a large, olive-colored pine cone. The outer skin, which is inedible, has medallions that overlap in a pattern, much like the armorlike skin of some reptiles. Cherimoyas are harvested while quite firm because they are easily bruised unless handled with utmost care. The tree-ripened fruit is superior in flavor, but unless you live in California or Peru, you'll probably never be offered tree-ripened cherimoyas.

The flesh of a ripe cherimoya is white or creamy in color. It is similar in texture to a baked apple or custard, hence the name custard apple. In flavor it tastes like a rather bland blend of pears, pineapples, and bananas. Although they are usually used as a table fruit, they are also used in fresh fruit salad. Adding a little fresh lime juice helps to perk up the rather bland flavor.

Eating cherimoyas with a spoon is a futile task (you spend more time spitting out the seeds than enjoying the fruit), and preparing them to add to a fruit salad requires much patience. The cherimoya has many large, black, inedible seeds embedded throughout its flesh. The best way to serve it is as a raw fruit.

Because of their fragility, they must be harvested and packed carefully in small, well-cushioned cartons.

FRESH DATES

The date is the fruit of a tall palm tree known as the date palm. Dates are the staff of life for people who dwell in the arid regions of North Africa and the Middle East. The date palm bears its first fruit in the fourth year and continues to do so, with little or no care, and under far from ideal growing conditions, for the next seventy-five years.

In the United States dates are grown commercially, on a fairly large scale, in the desert areas of California and Arizona, but some are also imported from Israel. The best-known variety is called the Deglet Noor. The largest and most costly variety is called the Medjoul.

Fresh dates used to be marketed with pits. In recent years the growers have found a method of removing the pits without mashing up the delicate fruit, which has resulted in greatly increased popularity.

No preparation is required. Serve them as you would dried apricots or prunes. They are excellent when used in baking. Look for moist, soft dates; if they are hard and dry, they won't be nearly as flavorful.

FRESH FIGS

Figs date back to antiquity and are believed to have originated in Asia Minor. They are a very important cash crop in Greece, Turkey, and Italy. Nearly all the figs grown commercially in the United States are produced in California, although Texas has a small crop that is usually sold to canners. Figs are also grown in many home gardens as far north as New York.

Figs are perishable and have a limited shelf life. They are fragile and have to be picked, packed, and shipped with utmost care. The California figs are in season from June through November. Those sold in areas any distance from California are usually sent to market via airplane. This combination of fragility and high cost of freight results in high price tags. The fruit is not only delicate to handle but also has a very delicate flavor. While they are not everyone's cup of tea, those who are fresh fig aficionados will pay almost any price to enjoy this unusual fruit.

There are several varieties of figs and they come primarily in two colors: light (also called green or white) and dark (also called black or purple). The best known light varieties are the Calmyrnas and the Kadotas. The best known dark varieties are the Black Missions (these were first planted by the monks in California) and the Brown Turkeys (named after the country and not the bird).

Most of the world's figs are not sold as a fresh fruit but are dried in the sun and shipped to the four corners of the earth as a dried fruit.

GUAVAS

Guavas, native to Mexico, are now grown extensively in Hawaii and on a much smaller scale in Florida. They are available from December through February.

The guava is a light-green- or yellow-colored fruit, about the size and shape of a plum. When tree-ripened, guavas have a pleasant strawberrylike flavor. When they are force-ripened, they are apt to be quite tart. Depending on the variety, the flesh color can range from white to dark pink. Numerous small seeds are embedded in the flesh near the heart of the fruit.

Guavas make a fine amber-colored jelly that was highly prized in Grandma's day. Of late, guava jelly seems to have gone out of style and is seldom found on supermarket shelves.

The best, most flavorful guavas are produced in Hawaii, but you'll have to visit the islands to sample their fine product. Due to stringent and wise law enforcement by the USDA, the guavas, which can be a host to the dreaded fruit fly, cannot be exported to the mainland unless they are fumigated. Unlike the Hawaiian papayas, the guavas are too delicate to survive the heat used as part of the fumigation process and they are never shipped or flown to the continental forty-eight states. Unfortunately, Hawaii is one of the fruit fly capitals of the world.

KIWIS

Kiwis, which used to be called Chinese gooseberries, originated in the Yangtze Valley of China. Early in the twentieth century, the Yang Tao, as the fruit is called in China, was transplanted to New Zealand, where it was renamed the kiwi. They called this odd-looking, fuzzy, brown fruit a kiwi because it resembled their funny-looking, fuzzy, brown kiwi bird. The New Zealand climate proved to be ideal for growing kiwis, and this fruit soon became one of that country's major exports. Other nations quickly climbed aboard the bandwagon and now kiwis are grown in many lands. Early in the 1960s the kiwi was introduced to Southern California, where it also thrived and is now a major cash crop.

Kiwis have a seven-month season. The California crop is in market from May through November and the New Zealand fruit is available from November through May. Since New Zealand in the Southern Hemisphere and California in the Northern Hemisphere have reverse seasons, the kiwi crops overlap, making the fruit available year-round.

On the surface, the egg-sized, egg-shaped, fuzzy, brown kiwi is one of our least attractive fresh fruits, but its drabness is only skin-deep. A cut kiwi reveals flesh that is an attractive lime-green in color containing hundreds of tiny edible seeds embedded in a geometric pattern.

The kiwi is juicy with a smooth texture and a tangy, tart-sweet taste. Tutti-frutti might be an accurate description of its flavor because it tastes like a blend of watermelon, strawberries, and grapes. While the kiwi blends well with other fresh fruits, it is at its flavor best when served on its own.

Kiwi juice is delicious, but it is too costly to serve solo as a beverage. However, when added to iced tea, fruit punch, or cocktails, it makes a refreshing drink and an interesting conversation piece.

As more and more people discover this unusual fruit, the demand for and the sales of kiwis show an annual increase. Not only has the kiwi been discovered by the consumer at the produce counter, it has also made significant inroads in bake shops and restaurants and with caterers.

This ever-increasing demand has been more than matched by increased production. Each year more tonnage is shipped to market and prices have gradually decreased as the supply has exceeded the demand. The kiwi is the only fresh fruit that is less costly today than it was twenty years ago. Although prices have eased, kiwis are still fairly expensive. They are usually sold by the piece and not by weight.

When purchasing kiwis, select fairly firm fruit and allow them about a week to ripen at room temperature. If you select already ripe fruit, they may be bruised. When the kiwi has a slight yield to gentle pressure, about as much give as a ripe plum, it is ready to eat.

The best way to prepare a kiwi is to cut it in half lengthwise, from stem end to blossom end. Once it is halved it is easy to pare off the inedible skin. If you try to peel the kiwi prior to halving it, it will be difficult to remove the skin without waste and mess. Once the kiwi is peeled, it is easy to slice the fruit. Glazed kiwi slices are especially attractive when used as a topping for pastry.

☐ WHEN TO BUY: *Available year-round*
☐ WHAT TO LOOK FOR: *Plump, firm fruit*
☐ HOW TO STORE: *Ripen at room temperature; refrigerate when ready to eat*

LOQUATS

The loquat, like the lychee, is of Chinese origin. Although popular with Asians, it has yet to make significant inroads in Western culture.

Loquats look like small, downy-skinned apricots, which they only resemble in shape and color. Unlike the apricot, which contains a single stone, the loquat has three or four pits. When loquats are ripe, they are fairly sweet and taste somewhat like a Royal Anne cherry. However, they have more pit than edible flesh. Their peak of season is mid-March to May. They are one of the earliest of the summer stone fruits to arrive on the scene. At one time this early debut added much to their desirability. Today, when so many fruits once limited in season are now available year-round, loquats are no longer prized for their early arrival. Unless you want to try them as a curio, they are usually too costly to be a good buy. Serve as a raw fruit.

LYCHEES

Lychees, which are also known as lychee nuts, are of Chinese origin and are highly prized throughout the Far East. About the size of a golf ball, the lychee has a rough, parchmentlike skin that is strawberry-colored when first picked and then fades to a dusty pink. Beneath this easily-peeled inedible skin is a round hard nut, also inedible, which is covered with a thin layer of light green flesh that is similar in flavor to a muscatel grape. The entire edible portion of the lychee is at most equal to two or three grapes. This minimum of edible yield when combined with the usually costly price tag makes the lychee one of our most expensive fruits. Serve it as a raw fruit.

Lychees are grown in Florida on a limited scale and the crop is usually shipped to large urban areas with sizable Asian populations.

Dried lychees, which taste like raisins, are available twelve months of the year. Both the fresh and the dried lychees are popular Oriental gifts.

The combination of a limited supply and great demand in Oriental neighborhoods makes it highly unlikely that you will find lychees in the average supermarket.

MANGOES

Mangoes are believed to have originated in India and Burma. They are a most flavorful and refreshing fruit that grow on huge trees, some of which attain a height and width of near fifty feet. To the people in the tropics, the mango tree plays a role similar to that of the apple tree in North America. Some claim that the forbidden fruit in the Garden of Eden was a mango rather than an apple, and its out-of-this-world flavor lends credence to the claim. I know of no other fruit, with the possible exception of a vine-ripened honeydew or a perfectly ripe pineapple, that is any sweeter or more fragrant than a ripe mango.

Mangoes are in season from January through September. The peak of season, which features the lowest prices and the most flavorful varieties, is May, June, and July. Most of our mangoes are imported from Mexico and Central America. Some are brought in from Haiti, and of late a few have been flown in, during the off-season, from Brazil. We also have a fair-sized crop that is produced in southern Florida. Hawaii grows mangoes, but as a rule its crop is less than sufficient to supply the demand in the islands. For the most part, attempts to grow mangoes in California have not borne marketable fruit.

Mangoes come in assorted varieties, sizes, colors, and shapes. They can be as small as a hen's egg and weigh a few ounces or as large as an ostrich egg and weigh about four pounds. Neither extreme of size is grown commercially. While there are countless varieties, only about half a dozen are sold in quantity in the United States. These can be broken down into two types: the Saigon mango and the Indian mango.

The Saigon type is represented by our imports from Haiti. These mangoes are fairly flat and kidney-shaped. As they ripen, they color up much like a banana, going from a dull dark green to light green and then to a dull yellow, which is why they are also called Banana mangoes. The ones we import early in the season from Haiti are not very good, but those that arrive at the tail end of the season are called Francines and are very good. Unfortunately, both the losers and the winners are as alike in appearance as two peas in a pod. The trick is to skip the Haitian mangoes until late in the summer.

Indian mangoes are imported from Central America and Mexico and are also grown in Florida. There are five major varieties—three are excellent, one is an also-ran, and one is bad news.

The one to avoid is a fairly flat, kidney-shaped, green-skinned mango with a red cheek, called an Oro. There is nothing golden about this variety except the color of its flesh. It is quite stringy, tastes like turpentine, and usually spots up and decays before it ripens. If you have ever bought an awful mango, it was probably an Oro. They are brought in from Mexico and should be turned back at the border by the USDA or the U.S. Customs Service. The only reason they sell is that they look quite good and are the first variety to hit the market.

The also-ran is our best-looking and best-selling mango. It is called the Tommy Atkins (which is the British counterpart to our G.I. Joe). These are shapely, smooth-skinned, and as pretty as a picture. The skin color is almost completely bright red. The flavor is fair, but the fruit tends to be stringy. Why is it our best seller if it isn't as flavorful as some of the other varieties? Because American consumers often make a choice by their eyeballs rather than their taste buds.

One of our winners is a variety called the Haden. It isn't very large in size and when ripe it is yellow in skin color with a red cheek. It is very fragrant and is our sweetest, tastiest mango. It has a fairly good texture, measured by its lack of fiber or strings.

The Kent variety is a fairly large, green-skinned mango that has a reddish cheek. It is sweet in flavor, has a smooth texture, and is fiber-free.

The Keitt variety is large, green in color, and may or may not have a slight touch of red. It is fairly sweet and has a very smooth, fiber-free texture. It also has a smaller seed than other mango varieties.

The Haitian fruit arrives as early as January. They are joined by the Oros from Mexico in February. Both are bad news and should be avoided until the Mexican Hadens arrive in April. From April through September it's clear sailing and good eating. During those six months, fine Central American, Mexican, and Florida Hadens, Kents, and Keitts are in the market. In September the Haitian Francine variety is also worth buying. The Tommy Atkins—at best only pretty good—are available from May through July.

Although Florida grows the same varieties that are grown in Mexico and Central America, our domestic fruit is never quite as sweet and juicy as the imports. While the warmer climates in Latin America are more ideal for the warmth-loving mango, the main reason why the Florida mangoes aren't as good is the time of harvest. The Mexican mangoes are left on the trees until they reach near-full maturity. When this fruit arrives at market it is either ready to eat or very close to it. The Florida mangoes are picked while hard as a rock and will take a week or more to ripen up at room temperature. The tree-ripened fruit has more flavor and fragrance.

While there can be some dispute as to whether the mango is the world's sweetest fruit, there is no argument that it is the sloppiest. This fruit wasn't designed for dainty eating. The combination of a very juicy flesh and a large, flat pit that is not freestone makes it imperative that you either wear a bib or have lots of paper towels handy when you tackle a mango.

In the tropics, where serving mangoes is a rule rather than an exception, they use silver mango forks with four long tines. A very ripe mango is skewered and the skin is scored with four lengthwise cuts. The skin is then peeled down like a banana and the fruit is eaten like an ice cream pop.

Another common way to eat mangoes in the tropics is to gently roll them on a table as you would to soften a hard lemon. When the pulp is almost liquid, make a small incision at the stem end and suck out the nectarlike pulp. This method can only be used if the mango is dead ripe.

If you perform this method gingerly and with great care, you won't need paper towels. However, I recommend that on your first try, you have the towels handy.

Mangoes can be cut into uneven slices and served solo or in combination with other tropical fruits. Hard green mangoes are used to make chutney.

It is very difficult to pinpoint and describe the flavor of a mango. The closest I can get is to say it has a tutti-frutti flavor. A ripe mango has the combined flavor of very ripe peaches, apricots, and pineapples. If you have yet to try a mango, you are in for a flavor treat. It isn't a taste that has to be acquired.

☐ WHEN TO BUY: *At peak May, June, and July*
☐ WHAT TO LOOK FOR: *Firm, unbruised fruit*
☐ HOW TO STORE: *Ripen at room temperature; preferable not to refrigerate*

OLIVES

Records show that olive trees were cultivated in ancient Egypt as long ago as 2000 B.C. In the Old Testament story of Noah, it was an olive twig that the dove returned to the ark to give word that the flood was starting to recede.

Ninety-five percent of the world's olives are grown in the Mediterranean basin, the two top producers being Spain and Italy. The bulk of the olive crop is pressed into olive oil. However, a fair portion is cured in brine or oil and put up in jars and barrels. Raw olives are as hard as a rock and aren't very perishable. Look for clear-skinned, unmarked olives.

In the United States nearly all of our commercially grown olives are produced in California. Most of the crop is allowed to ripen, then put into tins and sold as ripe black olives. A small percentage of the crop is shipped to market as fresh olives and usually purchased by retailers who have stores in predominantly Italian neighborhoods.

Raw fresh olives are inedible; they have a vile flavor. But when they are cured in brine or vinegar and flavored with garlic by someone with know-how, they are very tasty. Unless you have a determined olive maker in your family, buy your olives in jars at the supermarket.

PAPAYAS

The Papaya, believed to have originated in Central America, is now an important fruit crop in all tropical areas of the world. It is most unusual. Botanically it is a berry that grows on a tall, treelike plant that reaches a height of twenty feet. The fruit is pear-shaped, but cuts and tastes like a cantaloupe. Like a melon, the countless, tiny, round black seeds are contained not in the flesh of the fruit but in a cavity. Papayas, often referred to as melons that grow on trees, are in season twelve months of the year.

Just as the pineapple originated in Brazil and did well when introduced on the islands in the Caribbean basin, but reached its finest hour when transplanted to Hawaii, the papaya followed the same path to success. Until after World War II, unless you visited the Hawaiian Islands, you could not enjoy the fine flavor of the Hawaiian pineapple or the Hawaiian papaya. Today, thanks to the speed of the jet plane, both items are available in most stateside areas.

Papayas grow like weeds in the tropics. Puerto Rico, the Bahamas, and Florida have varieties that attain great size, some as big as a football. Hawaiian papayas are carefully tended and nourished. Almost all are of uniform size and are identical in appearance. Usually a light green in color when harvested, they color up to a golden yellow as they ripen, following the same changing color pattern as bananas. Nearly all the papayas sold in North America are grown in Hawaii and arrive via air freight.

The main Hawaiian variety is called Solo, but there is nothing solo about the way they grow. I have seen hundreds on a single papaya tree. The Solo is a yellow-fleshed variety. Recently there has been an increased supply of an orange-pink-fleshed variety called the Sunrise. Both varieties are equal in flavor and texture and will ripen at room temperature.

I am told that the black papaya seeds are edible if they are ground in a blender until they resemble coarsely ground black pepper and added to salad dressing, but I find their appearance far from inviting.

Fully ripened papayas are golden yellow in color. When they are green in color they are not mature and lack flavor. The best way to prepare them is to cut them lengthwise, from stem end to blossom end, and scoop out the numerous black seeds. Then add a few drops of fresh lime or lemon juice to jazz up the rather bland but sweet flavor. A ripe papaya will have a flavor and texture very similar to that of a ripe cantaloupe.

Due to the great distance that papayas have to travel, they are usually quite costly. They are not a good buy when the usually less costly and more flavorful cantaloupes are in season, but when melons are out of season—usually in December and January—papayas can be an excellent substitute. Although available twelve months of the year, they are in short supply and at yearly high prices from March 15 to May 15.

☐ WHEN TO BUY: *Available year-round*
☐ WHAT TO LOOK FOR: *Firm, pale green, or pale yellow fruit*
☐ HOW TO STORE: *Ripen at room temperature; preferable not to refrigerate*

PASSION FRUIT (GRANADILLA)

Passion fruit grows on a climbing vine that is native to Brazil. It is now widely cultivated in tropical and subtropical regions primarily as an ornamental vine that has a very attractive blossom (passionflower). In recent years New Zealanders have been trying to create a market for this exotic-looking fruit, but as yet their efforts have met with little success.

The fruit is usually purple in color, but some from New Zealand are a beautiful combination of red and gold. It is about the size and shape of an egg. Beneath the smooth inedible skin is a yellow pulp that is honeycombed with small black seeds. When the fruit is ripe and soft enough to eat with a spoon, the flesh is fairly sweet. However, the seeds are a nuisance.

Despite its rather sexy title, the passion fruit has no aphrodisiac qualities. To the contrary, it got its name from early missionaries to South America who perceived symbols of Christ's passion (the Crucifixion) in the various components of the flower.

PERSIMMONS

The persimmon, a native of ancient China and Japan and very popular in the Orient, is a most unusual fruit. Not akin to any other fresh fruit, it is related to the ebony tree. Just as ebony wood is highly prized for making furniture, persimmon wood is highly prized for making heads for golf clubs.

All of our commercially grown persimmons are produced in California because the fruit of the persimmon trees that grow wild in the Deep South are too small and fragile to have commercial value. The two varieties grown in California are the Hachiya and the Fuyu, both of which originated in Japan. Persimmons have a rather short season; they are in the market from October to January and a few are brought in from Chile in the spring.

The Hachiya persimmon is one of our most colorful and shapely fruits. It looks like a shiny, deep-orange-colored, acorn-shaped tomato. If allowed to ripen fully, there is no fresh fruit that is sweeter than a Hachiya. However, if it is eaten prior to reaching full ripeness, it is without doubt our worst-tasting fruit.

When purchasing Hachiyas, select those that are very firm and colorful. Don't buy those that have started to ripen up because they may be bruised. Allow the persimmon about three to four days to ripen at room temperature. When you are convinced that the persimmon is ripe enough to use, wait another day or two. When you are sure that it is overripe and ready for the garbage can, then and only then will it be ready to be eaten. By that time the once beautiful persimmon will have shriveled and lost color. Unless the skin looks like a blister, it isn't quite ripe. The skin of the persimmon is similar to that of a tomato. It may or may not be eaten. The edible part of the fruit is seedless. Serve it as a raw fruit.

Don't ever bite into a firm Hachiya. I tried it back when I was a youngster, some fifty years ago, and I can still remember vividly the unpleasant experience as if it happened yesterday. An unripe persimmon is more astringent than an equal amount of alum. Even if you drink a gallon of water after sampling the firm persimmon, it won't wash away the cottonlike pucker in your mouth.

There is an old wives' tale that if you freeze a rock-hard unripe persimmon, it will be dead ripe when it thaws out. Incredibly, this is true. If you put a hard Hachiya in the freezer and remove it when it is frozen solid, it will magically transform and be ripe, juicy, and not astringent when thawed.

Most of the persimmons grown in California are of the Hachiya variety, but they also grow a limited supply there that are called Fuyus. The Fuyus aren't nearly as large, pretty, or shapely as the Hachiyas, nor are they as sweet and juicy when ripe. However, they do have one big advantage: If eaten while still firm, they are not astringent.

☐ WHEN TO BUY: *At peak November and December*
☐ WHAT TO LOOK FOR: *Firm, colorful fruit*
☐ HOW TO STORE: *Ripen at room temperature; refrigerate when ready to eat (very soft to the touch)*

PINEAPPLES

The pineapple is a tropical fruit that is believed to have originated in Brazil and is now grown commercially in all tropical areas with a similar climate and soil. Hawaii, Taiwan, the Philippines, Mexico, and Central America are major producing areas.

The pineapple is highly prized as both a fresh and a canned fruit and canned pineapple juice is very popular. The Hawaiians were the first to can pineapples, which used to be their top industry before the advent of the jet age; now tourism is number one. Today, much of our canned pineapple is processed in Taiwan and the Philippines due to the much lower labor costs in the Far East.

Pineapples are not grown in the continental United States. They are imported from Mexico and Central America as well as Hawaii. On rare occasions a few are also flown in from Africa's Ivory Coast. Until a few years ago we received a fair supply of pineapples from Puerto Rico, but of late they have stopped shipping fresh pineapples to the mainland. Cuba too was once an important source.

There are several varieties of pineapple. Those grown in Hawaii and most of those grown in Mexico and Central America are of a variety called the Smooth Cayenne. Those grown in Cuba and Puerto Rico are usually of the Red Spanish variety. All varieties of pineapple are sweet and juicy if allowed to ripen fully before being harvested. Once the fruit is severed from the plant, it ripens no further. If picked after having reached full maturity and rushed to market without exposure to excess heat or chill, the pineapples will have a high sugar content and a juicy texture. If picked when immature, they will be woody in texture and not very sweet. If they have been chilled, they will cut black. If they have been overheated in transit, they will be very soft and possibly have spots of decay.

While the pineapple is native to Latin America, it did not reach flavor peak until it was transplanted to Hawaii. The pineapple thrives in Hawaii's climate and soil. Prior to World War II, those of us who had never visited Hawaii were told by tourists who had made the sea voyage to the islands about the marvelous flavor and size of the Hawaiian pineapples. The returning servicemen who had served in the Pacific repeated the story.

However, it wasn't until jets became common transport that Hawaiian pines became available in all of our larger cities. They arrive now at your favorite market no more than two or three days after they have been harvested. The Hawaiian pineapples that you purchase locally are no different in flavor and texture from those sold or served in Hawaii. They do, however, carry higher price tags because of the high cost of air freight.

Not all pineapples are flown in via jet from Hawaii. Most sold in the United States arrive via truck or boat from Latin America. These Latin American pines look very much like the ones from Hawaii because both areas grow the same variety. They are lower in price but usually not as sweet and juicy as the Hawaiian pineapples. As a rule, the Latin American pines are picked while still too green to reach full sugar content. At best they are pretty good, but too often they cut woody and aren't sweet. Often during the winter months they cut black. Purchasing a Latin American pineapple is a gamble and there are at least as many losers as there are winners. Purchasing a Hawaiian jet pineapple, even though it is more costly, is a far safer bet. If you select one that is firm and unbruised, it's almost a sure thing.

The two biggest shippers of Hawaiian pineapples are Castle & Cooke, which uses the Dole label, and Del Monte, which uses the Del Monte label. Castle & Cooke has the lion's share of the market. Both firms attach paper name tags to the crown of each pineapple and both firms ship (actually fly) superb fruit of comparable quality at comparable prices.

These same two firms are the major shippers of Central American pineapples. Castle & Cooke (Dole) grows them in Honduras and Del Monte in Costa Rica. They also attach paper labels to the crowns of these pineapples. If a pineapple doesn't have a name tag, it is probably a product of Mexico. Unfortunately, not only do the pineapples from Hawaii and Latin America look alike, the name tags are also deceptively similar. Millions of consumers have purchased Latin American pineapples, assuming they came from Hawaii because of the look-alike name tags. If you don't read the labels carefully, you won't get the real thing.

The Dole name tag on the pineapples from Hawaii has the Dole logo, spells out ROYAL HAWAIIAN—JET FRESH, and has a picture of a 747 jet plane. Any other name tag that has the Dole logo or reads DOLE—PREMIUM but makes no mention of the source is a Honduran pineapple.

The Del Monte Hawaiian pineapples bear the familiar Del Monte name tags that also say: JET FRESH FROM HAWAII. Their Central American pineapples have look-alike labels that read DEL MONTE PINEAPPLE. Here too no mention is made of the source on the front of the label. But Costa Rica appears in small print on the back of the label.

In some areas, retailers display or advertise the Latin American fruit as Hawaiian pineapples. Although this deception is not condoned by the growers, labeling their fruit as to source could easily put an end to this practice.

One of the latest gadgets at the produce counter is a machine that removes the skin of and cores pineapples. While this may save time, it also wastes pineapple because the die cannot be adjusted to the size of each fruit. The most economical way to prepare a fresh pineapple is first to twist off the green crown. Then lay it on its side on a cutting board. Slice the pineapple as you would a loaf of bread in one-inch slices. Pare each slice as you would an apple. This method won't take much time and very little pineapple will be wasted.

When selecting pineapples, first check out the name tag to determine the source of the fruit. Then pick the largest one in the pile that is firm and shows some color. The Hawaiian growers claim that the shell color is not a clue to ripeness, and the pineapples from Hawaii with green shells are picked at maximum ripeness, but I find that those that have a trace of gold or orange color are sweeter and juicier. Occasionally you'll see a pineapple that has a twin crown. While this extra foliage is very attractive, don't buy a double-topped pineapple. If it has two tops it will also have two cores. Pulling the leaves from the crown to determine the ripeness of a pineapple is an old wives' tale and is as valid as kicking the tire of a used car.

The flesh of a fully ripe pineapple will appear to be glossy and wet. This is a plus and not a minus. It is a sign of high sugar content and full ripeness.

Pineapples should never be stored in a place where the temperature is below 50° F. Since refrigerators have a range between 34° F and 38° F, they are too cold for pineapples. When pines are kept under refrigeration for more than a few days, they may cut black.

☐ WHEN TO BUY: *Available year-round*
☐ WHAT TO LOOK FOR: *Firm fruit with a trace of orange color; "Hawaii" "Jet Fresh" on the label*
☐ HOW TO STORE: *Hawaiian pines are ready to eat when purchased; never store uncut fruit in the refrigerator*

POMEGRANATES

The pomegranate dates back to ancient Persia. It is a most unusual fruit that resembles no other in structure. While it is far from a household word in most areas of North America, it is very popular with people of eastern Mediterranean, Near Eastern, and Far Eastern backgrounds. Our entire pomegranate crop is produced in California, and much of this crop is exported. The pomegranate has a short season, from October to January. Although they are also grown in the Southern Hemisphere, they are seldom imported during our off-season.

Pomegranates grow on trees, and the round, bright, red-skinned fruits look like Christmas tree ornaments when they reach full color. There are also some golden-yellow varieties, but these are not grown commercially. Most of the ones grown in California are of a variety that is modestly called Wonderful. The skin of the pomegranate is leathery and parchmentlike in texture. The flesh is honeycombed with hundreds of ruby-red kernels, each chock full of juice and containing one rather firm but small, chewable seed. As with grape seeds, you can either chew them or discard them, but they are quite chewy. While some may consider the seeds a nuisance, the fruit is very tasty and very juicy. It has a most refreshing tart-sweet flavor. It is almost as juicy as a citrus fruit and is often squeezed to yield its red juice, which is used to make grenadine syrup. This claret-colored juice is almost indelible. Long before the introduction of paper and ink, the ancient Egyptians used pomegranate juice to write on papyrus. They also used it to dye fabrics.

Most adults don't have the time it takes to eat a pomegranate. However, to children the pomegranate is both a fruit and a toy. They are fascinated by the jewellike kernels. Since the juice is almost indelible, it can create a most colorful mess if given to an unattended youngster. Pomegranates need not be refrigerated, but refrigeration won't hurt and will extend their shelf life.

☐ WHEN TO BUY: *At peak October and November*
☐ WHAT TO LOOK FOR: *Firm, colorful, unbruised fruit*
☐ HOW TO STORE: *No refrigeration required*

PRICKLY PEARS

The prickly pear is also called the cactus pear and the Indian fig. It is actually neither a pear nor a fig but the fruit of a variety of cactus. It originated in Mexico and was introduced in Spain by the returning explorers. This unusual fruit is now grown, and even grows wild, in areas that border the Mediterranean Sea. In the United States it is commercially grown in limited quantities in Southern California.

Depending on the variety, this pear- or fig-shaped fruit comes in many colors: pale yellow, orange, pink-magenta, and blood red. The latter color dominates in the fruit grown in California. Although it is the product of a plant that requires little moisture and thrives in very arid areas, the flesh of the prickly pear is very juicy. The flavor is a refreshing combination of tart and sweet.

The fruit is colorful, flavorful, and juicy, but it has two severe drawbacks: The pulp is full of small, hard seeds and the outer skin is covered with small, thin, very sharp barbs.

The prickly pear is one of our most seedy fruits. While it is claimed that those seeds are edible, most people find them too hard to chew. Serve the prickly pear as a fresh fruit. Cut off the top and bottom, slit it from end to end, and peel off the inedible skin.

However, the problem with the thorns has been corrected. The prickly pear is no longer hard to handle because the sharp, thin barbs are removed by singeing and brushing prior to shipment of the fruit to market. To accentuate the lack of barbs, the growers now market what used to be called prickly pears as cactus pears.

Way back when I first started in the produce business, we used to dread the arrival of the prickly pears. No matter how carefully or gingerly we handled the fruit, we usually wound up with many of these sharp, very difficult to remove barbs embedded in our hands. Now, if growers could only remove the seeds, they'd really have a winner.

☐ WHEN TO BUY: *At peak from July through December*
☐ WHAT TO LOOK FOR: *Larger-sized, red-fleshed fruit*
☐ HOW TO STORE: *Ripen at room temperature, then refrigerate*

Berries

BLACKBERRIES

·

BLUEBERRIES

·

CRANBERRIES

·

CURRANTS

·

GOOSEBERRIES

·

MULBERRIES

·

RASPBERRIES

·

STRAWBERRIES

Berries are small succulent fruits that grow on vines, canes, or small bushes. They usually have seeds. In some types the seeds are hard, fairly large, and very obvious, and in others the seeds are minute and go unnoticed. While some berries are more hardy than others, all varieties are perishable and should be stored under refrigeration.

BLACKBERRIES

The blackberry, like the raspberry, is a bramble fruit. It is a member of the heath family. One of the theories used by scientists to support the belief that North America and Europe were originally one land mass is the fact that wild blackberries are native to both continents yet the distance across the Atlantic is too wide for the seeds to have been scattered by the birds or by wind.

In climates that are ideal for blackberries to flourish, they will grow like weeds if left unchecked. Not too long ago, unattended blackberries could be found growing at random along the roadsides of rural areas in our northern states. In some areas they are known as dew berries.

Today, blackberries are grown commercially on a large scale in the Pacific Northwest, Michigan, and New Jersey. They are also produced in many other states on a much smaller scale. Most of the blackberry crop is sold to commercial processors and made into jams, jellies, syrups, and liquors.

There are two types of blackberries: those that grow on arched canes and those that grow along the ground. Both types are similar in flavor, color, and texture. There are some hybrid blackberries that are grown on the West Coast. These are much larger than the original black-berries, and usually lighter in color and not quite as acidy. Some of these hybrid varieties include boysenberries, loganberries, and youngberries. These are more fragile than the true blacks. They are as delicate as rasp-berries and their seeds aren't quite as hard as those of true blackberries. Some hybrids are almost seedless.

Blackberries are in season from May to October, with the peak in June and July. During the winter months a very limited supply is flown in from the Southern Hemisphere.

While the true blackberries aren't quite as fragile as raspberries or the hybrid blackberries, they still have to be treated as highly perishable and should be kept under constant refrigeration. Even if they are in per-fect condition at the time of purchase, they have a shelf life of two to three days at best. Very often the blackberries offered in the marketplace are far from perfect.

So check them out carefully prior to making a purchase, especially those that are sold in full-pint baskets as opposed to shallow half-pint trays. If they are soft, wet, stuck together, or show any trace of mildew, you'll probably have to discard more than you can salvage.

While the hybrids can range in color from violet to a reddish pur-ple, the ideal color for the true blackberries is a jet black. As these black-berries ripen on the cane they go from green to red and only to black when fully ripe. The blacker the blackberry, the better the flavor.

BLUEBERRIES

Some botanists estimate that wild blueberries flourished as far back as 10,000 B.C. They are native to the northern regions of the Northern Hemisphere. Blueberries were staples in the diets of American Indians and Eskimos prior to the arrival of the explorers from the Old World. A sim-ilar type of berry grown in Britain was called the whortleberry.

The annual commercial North American blueberry crop (both the wild and the cultivated blues) comes close to 200 million pounds each year. That's 95 percent of the world's total production. While blueberries are gaining popularity in other parts of the world, nothing quite compares with the love affair that Canadians and Americans have with the colorful, flavorful blueberry.

Until early in the twentieth century there was only one type of blueberry: the wild blues that were used by the Indians and dated back to prehistoric times. These grow on low bushes, six to eighteen inches in height. Due to the short stature of the bush and the very small size of the berry, these wild blues can't be harvested by machine. They have to be handpicked or gathered with a wooden rake after the bush has been shaken by hand. Gathering these wild blueberries on a commercial scale is both slow and costly.

Today's low-bush blues are still known as wild blueberries even though they are grown commercially. Most of them are grown on privately held land and are produced on fenced-in acreage. Every other year after harvest, the fields are burned down to the ground, eliminating the need for pruning or weed control. The only thing wild about these wild blueberries is the price tag. Those plentiful wild blues that were free for the taking by the Indians and early settlers are now usually quite costly. Most of them never reach the marketplace as a fresh fruit. They are sold to commercial processors who freeze them or put them up in tins.

It was not until the early 1900s that the wild blueberry was tamed. The first cultivated blueberry made its debut in 1906 as a result of experiments in New Jersey conducted by the U.S. Department of Agriculture. These new cultivated blues were an instant success because they were much larger than the wild blue, had a more attractive color, and had smaller, less noticeable seeds. The cultivated blues have been improved by a constant stream of new varieties.

The annual tonnage of the cultivated blues has increased geometrically. In less than a century the annual tonnage has risen to close to 100 million pounds. This incredible and swift growth has never been matched by any other agricultural item.

Unlike their low-bush antecedents, the cultivated blueberries grow on bushes that may reach heights of more than eight feet. The greater height of the bushes and the much greater size of the berries permit them to be harvested mechanically.

The cultivated blue is not only three to four times larger than the wild blueberry, it is also more colorful. The wild blueberries are a dark (almost black) blue in color. The cultivated blues are a much paler powdery blue. Which type is more flavorful? That argument has been going on for years and has yet to be resolved.

The cultivated blue has all but replaced the wild blue in the market. Cultivated blues are almost flawless dessert fruits. Not only do they have fine flavor, they require little or no preparation other than a cold water rinse. They are ready to serve at the time of purchase. They don't have to be peeled, pitted, sliced, or cored, nor do they have to be further ripened or aged. While they do have seeds, those seeds are so minute and tender that they go unnoticed. With all these attributes, the cultivated blueberry comes very close to being a perfect fruit.

If the cultivated blues have one fault, it is the shortness of their season. They are available less than five months a year. Only cherries and apricots have shorter seasons. Cultivated blues arrive in May and wind up in September. There are a few blues flown in during the winter from the Southern Hemisphere, but these imported blues are usually very expensive because of the high cost of air freight.

The season for the cultivated blue is kicked off each year early in May by a trickle of a new variety of berry called Rabbit Eyes that are grown in northern Florida and Georgia. The tonnage is still very light, but the shippers plan to increase their acreage. Late in May the season gets into high gear with the arrival of fairly heavy supplies from North Carolina. In June the huge New Jersey crop comes on line. Oregon starts to ship in July. Michigan, Massachusetts, Washington State, and British Columbia blues arrive in August and wind up the season in late September.

Cultivated blues are usually marketed at retail in paperboard pint baskets. The net contents of each pint is fourteen to fifteen ounces. A pint of blues yields four servings, but a pint of strawberries yields but two to three.

Color is one of the hallmarks of quality and offers an effective way to judge their worth. The best blues have a powdery light blue color. They appear to have been dusted with a waxy powder. This wax was put there by Mother Nature and serves to protect the blues from the direct rays of the sun. This wax coating, which also appears on several other fresh fruits and vegetables, is called bloom.

The bloom is a clue to freshness. Although it won't rinse off in cold water, it gradually fades away within about a week of harvest. As the bloom disappears, the color of the berry gradually changes from light blue to almost black. Dark black blueberries aren't as fresh, firm, or flavorful as the light-colored blues.

Firmness and dryness go hand-in-hand, as do softness and wetness. Soft blues are on the verge of breaking down and are starting to decay. Always check the bottom of the paperboard pint baskets for telltale stains or leaking. Never purchase leaky blueberries. This same advice also applies for blackberries, raspberries, and strawberries.

In most of the other fresh fruits (with the exception of Bing cherries), size has little bearing on flavor and texture. Even though the larger-sized fruits are usually more costly, they are not any more flavorful than smaller, less costly fruit of comparable quality. However, when purchasing cultivated blues, the bigger the berry, the better the flavor. Pay the premium price and buy the largest blues available.

☐ WHEN TO BUY: *At peak June, July, and August*
☐ WHAT TO LOOK FOR: *Large, dry, plump, powdery fruit that are light blue in color*
☐ HOW TO STORE: *Refrigerate immediately*

CRANBERRIES

The cranberry, like the blackberry, is a member of the heath family. Native to North America, wild cranberries—as well as wild blackberries, blueberries, and Concord grapes—flourished in the area that we now know as New England long before anyone came over on the *Mayflower*.

Today, growing cranberries is a big, big business. They no longer grow wild but are cultivated by hundreds of individual farmers on privately held land. Today's cranberries are mechanically planted, irrigated, weeded, sprayed, and harvested. Even the pollenization is performed by bees that are rented. The annual crop exceeds 200 million pounds.

The major growing areas for cranberries are Massachusetts, New Jersey, Wisconsin, Oregon, and Washington State. The first fresh cranberries arrive in market early in October and the season usually winds up by the end of the year. As yet no fresh cranberries are brought in from the Southern Hemisphere.

There are four major varieties of cranberries. The Early Blacks and the Late Howes are grown in Massachusetts and New Jersey. The Searles are produced in Wisconsin, and the McFarlens, in the Pacific Northwest. All four varieties have fine color and excellent flavor. The Early Blacks are somewhat smaller in size and darker than the other varieties. While they have a slightly better flavor, they are not nearly as firm and hardy. The Early Blacks are also fairly perishable and have a rather short shelf life. They don't store well, even under refrigeration. However, the later varieties will last for several weeks in a refrigerator.

Cranberry sauce is an all-American favorite but has never quite caught on in Europe. We used to serve cranberry sauce only on Thanksgiving and Christmas, along with the traditional holiday turkey. Today it is served year-round as a garnish to other poultry and meat dishes. Very few people bother making fresh cranberry sauce at home, but substitute the sauce that comes in cans. Canned cranberry sauce is a fine product, but it doesn't quite match the color and the flavor of fresh sauce.

Another easily prepared taste treat is homemade fresh orange-cranberry relish. All you have to do is to grind up raw cranberries with a couple of fresh California oranges (peel and all) and sweeten to taste. The end product is delicious and inexpensive. Raw cranberries make an excellent substitute for raisins in baking breads, muffins, and cakes.

Nearly all of the huge annual cranberry crop is used either to make bottled cranberry juice cocktail or canned cranberry sauce. Only about 10 percent of the crop is sold as a fresh fruit.

While there are hundreds of farmers who grow cranberries for a livelihood, nearly all of the crop is marketed or processed by a single cooperative organization—the Ocean Spray Growers, Inc., of Plymouth, Massachusetts. Ocean Spray, one of the largest, and probably the most successful, agricultural co-ops in the nation, has very successfully marketed the cranberry juice cocktail. Thanks to the popularity of cranberry juice, there is seldom, if ever, an oversupply of fresh cranberries.

In the past twenty years the retail price of fresh cranberries has almost tripled while the net weight of a package has been cut from sixteen to twelve ounces. However, even at the higher prices fresh cranberries are still worth buying.

When shopping for cranberries, look for those that are dry, plump, firm, and colorful. If the berries are wrinkled, soft, wet, or go "squoosh" when you apply gentle pressure, they are overripe and should not be purchased. Fresh cranberries are in market only about three months of the year but can be available year-round since they freeze beautifully. Just put the package into a freezer bag and freeze. Don't wait until the tail end of the season to freeze the cranberries. Freeze them just around Thanksgiving, when they are at the peak of season; there will be less chance of getting overripe cranberries.

□ WHEN TO BUY: *At peak October and November*
□ WHAT TO LOOK FOR: *Firm, dry, dark berries*
□ HOW TO STORE: *Refrigerate immediately*

CURRANTS

The currant is a tart berry, about the size of a tiny seed pearl, that grows in clusters. In the United States we grow only red currants, but in Britain and France they grow black as well as red. The black currant may be host to a fungus that severely harms pine trees, which is why it is illegal to import black currant shrubs from overseas.

Currants are primarily used in making jams, jellies, and syrups. (They are a bit tart to use as a raw fruit.) Back in Grandma's day, when homemade jellies and preserves were popular, a fair amount of fresh red currants were sold in retail stores every spring. Today, they have all but disappeared from the marketplace. The commercial red currant jelly has also all but disappeared from the supermarket shelves. However, in Europe currant preserves are best sellers.

GOOSEBERRIES

Gooseberries are very tart berries that grow on a spiny bush. They look like light green translucent seedless grapes and when they are very fresh they are covered with fuzz. As they age and ripen, the pale-green hue changes to bronze and the fuzz disappears. Even when the green gooseberries are cooked, they lose color and turn brown. Gooseberries are too tart to use as raw fruit, but they make a superb jam.

Like currants, which aren't nearly as tart, gooseberries have all but disappeared from our marketplace, although they are highly prized in Europe. The British make a fine gooseberry jam and the French use them to make a sauce that is traditionally served with baked mackerel.

MULBERRIES

Mulberries grow on trees and bushes and come in two colors: black and ivory. The black mulberries look like blackberries and the ivory mulberries look like albino raspberries. Both types are quite flavorful and practically seedless. However, they are far too fragile to ship to market.

Mulberry trees are a fine addition to a home garden because not only do they bear fruit with little or no care, they also attract wild birds, especially doves.

While mulberries play no significant role in the United States as a cash crop, they are an important fruit in the Middle East. People of that region not only enjoy them as a fresh fruit but dry them in the sun and use them the way we use raisins.

RASPBERRIES

Wild raspberries, which once flourished throughout North America, have all but disappeared. Now, only cultivated raspberries are available in limited supply and they are usually very costly.

The raspberry and its cousin the blackberry are bramble fruits and are members of the rose family. But unlike blackberries, which are fairly firm and solid, raspberries are fragile and hollow. The structure of the raspberry makes it almost too delicate to handle and it has to be picked and packaged with utmost care to prevent crushing or bruising. Raspberries are usually marketed in shallow half-pint containers and have to be rushed to market. At best they have a shelf life of a day or two after they reach the retail market. This need for extra care assures that raspberries always carry a high price tag.

If you live north of the Mason-Dixon Line, you have access to native-grown raspberries three to four weeks each year and they probably sell at moderate prices. When the locally grown raspberries are not available in the Midwest and Northeast, these areas are supplied with raspberries flown in from California. California berries are available throughout the summer months and in early fall. In late fall they are joined by raspberries from our Pacific Northwest and British Columbia.

There are three schools of thought as to desirability of fresh raspberries. Most people can take them or leave them but usually leave them because of the high price tags. Some dislike them because of their seeds. However, those who adore their delicate flavor are willing to pay a premium price for fresh raspberries. In New York City in mid-winter they can sell for as much as seven to eight dollars per half-pint (about two servings).

While 99 percent of commercially grown raspberries are the traditional red color, there are also some black-, purple-, and apricot-colored raspberries. These unusual-colored berries are similar in flavor and texture to the more familiar red ones. The off-colored raspberries are usually grown in home gardens and are seldom sold in retail food stores.

Since the raspberry is both the most costly and the most fragile berry, special care should be used when making purchases. Avoid those that are soft, wet, or show any trace of mildew. If the raspberries are stuck together or if there is any trace of stain or leakage at the bottom of the half-pint container, you'll probably be able to salvage less than half the berries. Don't assume that a high price tag will be accompanied by high quality. If the raspberries aren't firm, colorful, and dry, pass them by.

Raspberries are as perishable as sweet cream. Ideally they should be used on the day of purchase. If not, store them under constant refrigeration.

☐ WHEN TO BUY: *At peak June, July, and August*
☐ WHAT TO LOOK FOR: *Firm, dry, colorful fruit*
☐ HOW TO STORE: *Refrigerate immediately*

STRAWBERRIES

Wild strawberries, which are also known as wood strawberries, have all but disappeared in the United States, though they are still commercially cultivated on a small scale in the Alpine areas of Europe and in season they grace the tables of fine restaurants in France. The tiny wild berries are long on flavor and fragrance.

The commercially grown strawberry we find in our retail stores are much larger and more abundant than the wild strawberries, but they aren't quite as fragrant and flavorful. However, we are continually producing new improved varieties and it is probably only a matter of time before we perfect strawberries that are as big as golf balls and will match the aroma and the taste of the wild berries.

Not only are today's berries superior to those that were available twenty or thirty years ago, they are in season for much longer periods of time. In areas north of the Mason-Dixon Line, the strawberry season once lasted only from early spring to the Fourth of July. Today, fine strawberries are available twelve months of the year.

Strawberries are grown in all fifty states, including Alaska. In most areas the strawberry season is about three months in duration. However, California strawberries are in season eleven months of the year and are even available in very limited supply during the month of December.

California is strawberry country. They grow more strawberries than anywhere else in the world and each year they add to the acreage used to grow the berries. With ideal climate and soil conditions, California produces more than 75 percent of the total United States crop. The berries thrive in sunny, coastal regions, where the days are warm but not torrid and the nights are cool and damp. California gets a tremendous yield per acre, seven times the average yield of the rest of the strawberry-growing areas in the country. They not only supply most of the North American market but also export to all the major cities of the world. In 1984 the size of the crop was a mind-boggling, record-breaking 560 million pounds—almost a 50 percent increase over the previous year.

While strawberries are shipped eleven months of the year, the peak of their season is from April 15 to July 15. During that period the finest berries of the year usually sell at the lowest prices of the year.

Florida is a very distant second to California in total production, and even at their best the Florida berries are no match for the best berries from California. However, beginning in December, Florida strawberries take up the slack for about six weeks, until the new crop of California berries comes on line. During the winter months we also import strawberries from New Zealand, Israel, and Mexico.

The quality of those berries air-freighted from New Zealand and Israel is quite good, but they are usually expensive. The Mexican berries arrive via truck and as a rule do not win any blue ribbons; however, they are usually not very costly.

The expensive, imported, out-of-season strawberries, even though they cost five dollars or more per pint, aren't nearly as good as the marvelous California berries that sell for well under a dollar a pint during May and June.

All strawberries are harvested by hand because they are too delicate to be picked by machine. Since stoop labor is both costly and hard to find, many of the smaller growers in all areas of the nation open up their fields to the public. This method of marketing is known as "U-Pick" and is also used to harvest other fresh fruits and vegetables. By selling direct to the consumer, the growers get a better price than they would by shipping their berries to the wholesale markets on consignment. The consumers not only pay less for the berries than they would in a retail store but get an extra bonus if they make the berry-picking a family outing. Children have a ball picking (and eating) the strawberries.

Select firm, dry, shapely, glossy, dark red berries that have fresh green caps. Strawberries ripen no further after harvest—what you see is what you get. Fine berries that meet the above criteria, if stored under refrigeration, can last for about a week without too much wear and tear. However, it is wiser to use them as soon as possible after purchase to enjoy top flavor and nutritional value.

Avoid strawberries (or any other berries) that are soft, wet, or show any trace of mildew. Also avoid those that are pale in color or have white shoulders. Until about ten years ago, the smaller- and medium-sized strawberries were more flavorful than the larger, usually more costly, berries. The bigger berries used to be hollow and less juicy. However, today's new large varieties have both size and flavor. Occasionally you may see very large strawberries with stringlike stems attached. These stem berries are a gourmet item that is highly overpriced. Unless you want these berries to dip in melted chocolate, or even in powdered sugar, at a dinner party, you can purchase equal or even better strawberries at less than half the price.

☐ WHEN TO BUY: *At peak April, May, and June*
☐ WHAT TO LOOK FOR: *Firm, dry, dark red berries; size has little bearing on quality*
☐ HOW TO STORE: *Refrigerate immediately*

Grapes

❧

BLUE SEEDLESS

·

WHITE (GREEN) SEEDLESS

·

RED SEEDLESS

·

BLUE GRAPES WITH SEEDS

·

GREEN GRAPES WITH SEEDS

·

RED GRAPES WITH SEEDS

A ccording to the Old Testament story, one of the first things Noah did after the ark landed was to plant a vineyard—proof that people have been enjoying grapes both as a fresh fruit and as wine since biblical times.

There are hundreds of varieties of grapes. Some are seedless but most have seeds, as did the original wild grapes. Grapes come in three skin colors: red, white (green), and blue. Some varieties have round berries, others have elongated ones. Some grapes are quite large, others are quite small. All grapes grow on vines and flourish in temperate climates in both the Northern and Southern Hemispheres. There are two basic types of grapes: the American and the European.

The American type, which is called the Labrusca, is native to North America. Leif Ericson is said to have called this continent Vinland because of its abundance of wild grapes. The earliest colonists found wild grapes thriving up and down the Atlantic Coast. The American grapes cannot be identified by shape, size, or color, but by their skin type. They have slip skins that are easily separated from the flesh of the grape. The seeds, however, are embedded and cling to the fleshy pulp.

The most common American variety is the blue-skinned Concord. There is also the red Catawba, the white (amber) Niagara, and the pink Delaware. Most of the Labrusca grapes are produced in the Northeast and Midwest, but some are also grown in the Pacific Northwest.

These slip-skin grapes have a fine aroma, a musky, semisweet flavor, and are pulpy in texture. Only a small portion of the slip-skin varieties is sold as table grapes. Nearly all of the supply is sold to commercial processors of jelly, jam, grape juice, and sacramental wine. The American grapes are in season for only two or three months during the fall and they are quite fragile. At one time, when putting up homemade jelly was in vogue, a fair supply of Concord grapes was sold to consumers. Today, only a limited amount reaches the retail marketplace. As yet, none of the American-type grapes are imported from the Southern Hemisphere in the off-season.

The European type, which is known as the Vinifera grape, originated in Asia Minor. Seeds of this type of grape, dating back some five thousand years, have been unearthed in archaeological digs.

The Vinifera grapes are also identified by their skin type, rather than by size or skin color. Instead of being slip skin, like the American grapes, their skin clings to the flesh of the fruit, and the seeds are fairly easily separated from the pulp. European grapes are produced in all of the grape-growing areas of the world. In the United States, California is European grape country and produces 98 percent of our table-grape crop. In a recent year the crop exceeded 50 million twenty-three-pound boxes of table grapes and the supply barely met the demand. The ever-growing popularity of table grapes is thanks to the efforts of the California Table Grape Commission, whose ad campaigns feature the flavor, nutritional value, and convenience of their product. In the past ten years our per capita consumption has almost tripled.

Fine table grapes are available at modest prices year-round because our supplies from California are augmented by imports in great quantity from Chile and in small quantity from South Africa. The California grape season extends from May through February. The Chilean season runs from December through May.

The seedless varieties are far more popular than those with seeds. Most Americans are either too busy or too lazy to mess with grapes that have seeds, but in Europe, where the pace isn't quite as rapid, these grapes are highly regarded. They are served as a dessert fruit along with cheese at the dinner table.

In North America, the grape is seldom used as a dessert fruit, but rather as a snack. In most retail markets grapes with seeds just don't sell, even if they are of better quality and are much lower priced than the seedless ones. Most of us are missing out on some very flavorful grapes because of our aversion to seeds.

Blue Seedless 🐛

The Black Beauty is the only blue-black seedless variety. It is also fairly new, but shows little promise because it is quite perishable and lacks flavor. It is very attractive and has semielongated blue berries. These arrive in market in June and wind up in July. This variety is long on looks but short on flavor and should be passed by.

White (Green) Seedless 🐛

The Perlettes are the first grapes of the season. They are grown in the desertlike areas of California and arrive in market in early May. They are round in shape and light green in color. They have a tender skin and when firm and fresh are very crisp and crunchy in texture. When they have high color (yellow rather than green), they are very sweet in flavor. When they are jade green in color, they are tart enough to curl your teeth. The California Perlettes are in market for only six to eight weeks. We also get Chilean Perlettes in December and January.

The Superior Seedless is a very tasty new variety. It is produced by only one grower who has been granted a patent. These grapes arrive late in May and have a short six-to-eight-week season. The berry is fairly elongated and has a very crisp texture and a tender skin. When it has high color (yellow), it is very sweet. Since they are produced by only one grower, the quality control is usually excellent. However, the demand exceeds the limited supply and Superior Seedless grapes usually command top prices.

The Thompson Seedless is America's, and probably the world's, favorite variety. In Europe they are known as Sultana grapes. The Thompson has a fairly good-sized, elongated berry and a tender skin. The ones from California arrive in June and wind up in November. In mid-January the Chilean Thompsons arrive and are available until the end of May.

Red Seedless 🐛

The Flame seedless is a new red-skinned seedless variety that shows great promise and may someday challenge the Thompson Seedless as our number one best-selling grape. It has only been around for a few years, but each year it captures a larger share of the grape market. The Flame seedless has medium-sized, very crisp and crunchy berries. The color is a pale red and the flavor is sweet and tasty. The skin is tender. The California Flame seedless arrive in mid-June and wind up in September. Flame Seedless are also imported from Chile during the winter months.

The Red Ruby Seedless is another new variety. While they have better color than the Red Flames, they have less size and flavor, and a tougher skin than the look-alike Flame seedless. They arrive in market in August, just as the Flames are winding up their season, and last until December. A few Red Rubies are also imported from Chile during the winter.

Blue Grapes with Seeds 🐛

The Exotic is a variety that is well named for its appearance but not for its flavor. It is the first blue-skinned variety to arrive in market, with a season from June to August. It features very large, colorful berries and a mild, not too sweet, flavor. It is fairly fragile. It looks like, and is often mistaken for, the later arriving but superior flavored Ribier variety.

The Ribier is the best blue-skinned variety. It features very large dark-blue berries. It has a crisp texture, a tender skin, and a sweet, full-bodied flavor. The Ribiers arrive in August and wind up in February. Note the August arrival date. Any blue grape purchased in June and July is of the less-flavored, look-alike Exotic variety. Ribiers are also imported from Chile during the winter months.

Green Grapes with Seeds 🐛

The Italia is an improved Muscatel grape (improved in looks but not in flavor). The Muscatel, which is the sweetest of all the varieties of table grapes, is very fragile and tends to turn amber in color when it reaches full maturity. This amber, almost brown, coloring doesn't appeal to the less knowledgeable shopper. The Italia is a type of Muscatel that retains its green skin color and has replaced the amber Muscatel as a table grape. It has large, round, pale-green berries and a sweet, full-bodied flavor. The Italias arrive in August and wind up in October. This variety is definitely worth trying even if you aren't too keen on grapes that have seeds. They are highly prized in Europe. As a rule they aren't imported from Chile because they are too fragile to make the long sea journey.

The Almeria is a variety that is being phased out. It has medium-sized light green and amber berries. The bunches are small and scraggly and the flavor is mild. The Almeria has been replaced by the more attractive, better-flavored Calmeria variety.

The Calmeria variety features elongated, slender, light green or golden berries. This gracefully shaped variety is sometimes called the Lady Finger grape. The Calmeria has a fine, mild, sweet flavor, a rather tough skin, and very few seeds. They arrive in October and last through February.

Tending vineyards requires intensive, backbreaking stoop labor, which cannot be performed by machine. The fragile table grapes must be carefully cultivated, harvested, and packed by hand. The growers, in combination with the agricultural schools, are constantly coming up with new varieties and improved know-how on producing superior grapes.

California also grows millions of tons of wine grapes, most of which are sold to wineries. A limited amount of these grapes are shipped to the larger cities to satisfy the demand in some ethnic areas. These wine grapes are not sold in supermarkets but by small neighborhood dealers. Making homemade wine is no game for a novice. Unless you have some expertise in winemaking, buy your wine in a liquor store. You'll have a more flavorful, less costly end product.

The raisin is our most important dried fruit and is produced by drying grapes in sunlight. Ninety-nine percent of our raisins are seedless, all of which are made from Thompson Seedless grapes. Golden-colored seedless raisins are nothing more than regular raisins that have been bleached with sulfur dioxide.

Cluster raisins that have seeds and are attached to the stems are made from Muscatel grapes. They used to be fairly popular in Grandma's era, but have all but disappeared from the marketplace.

Those tiny Zante currants that are used by bakers aren't really currants. They are tiny raisins made from the petite seedless black Corinth grapes.

When shopping for table grapes it is imperative to remember that the fruit doesn't ripen any further or improve in flavor after it has been severed from the vine. What you see, or better yet what you taste, if you are permitted to sample the grapes at the time of purchase, is what you get. The quicker you use them the better—as they age they lose crispness and flavor.

Look for firm, plump, colorful, dry berries that are firmly attached to pliable green stems. A professional buyer checks out the freshness of the grapes by examining the amount of bloom on the berry. As discussed earlier, bloom is the name given to the waxy, powderlike coating applied by Mother Nature to protect the fruit from the direct rays of the sun. This coating is more obvious on the darker-colored grapes, but it is also present, though not as easily detected, on the light-colored varieties. The heavier the bloom, the fresher the grape. As the grape starts to age and break down (after one to two weeks), the bloom disappears. Color is very important, especially in the green varieties. The greener the grape, the lower the sugar content. The yellower the grape, the higher the sugar content. Red varieties are at their best when the berries are predominately of high color. The darker the blue grapes, the better the quality.

Table grapes look fine as a centerpiece but will start to break down at room temperature. They must be stored in your refrigerator immediately after purchase and kept there until just prior to use. The crispness and the flavor of the grapes are enhanced when they are fully chilled.

□ WHEN TO BUY: *At peak (from United States): June through November*
　At peak (from Chile): February through May
□ WHAT TO LOOK FOR: *Firm, plump, dry fruit*
□ HOW TO STORE: *Refrigerate immediately*

Red Grapes with Seeds ❧

The Cardinal is a cherry-red-skinned variety that is as pretty as a picture. It was produced by crossing a Ribier grape with a Tokay. The Cardinal is the first dark-skinned variety (red or blue) to arrive in the market. It makes its debut in May and is usually gone by mid-July. It has only fair flavor and tough skin and is a fragile variety. It isn't nearly as flavorful as the Tokay or Queen varieties.

The Flame Tokay is the parent of the Flame seedless, and until its offspring made its debut, it was the best of the red-skinned varieties. It is still a superb red grape whose only shortcoming is a few seeds. The Tokay has a fairly large, pale-red berry with a tender skin, an excellent sweet flavor, and a crunchy texture. The Tokays arrive in July and wind up at the end of October. A few are imported from Chile during the winter months. This is one of the few table grape varieties (the others are the Thompson Seedless and the Muscatels) with a high enough sugar content to be used in making wine.

The red Queen is a fine new variety that is still in limited supply. It has the largest berry of any of the red-skinned varieties. It is colorful, tender-skinned, and crunchy in texture, and has a fine flavor. The Queens have a rather short season, available only in August and September. This variety has not yet been produced in Chile.

The Red Malaga is a variety that is gradually disappearing. Its cousin, the white (green) Malaga, has already been phased out. The Red Malaga has firm, medium-sized berries. The bunches are small and scraggly and the flavor is very mild.

The Emperor variety is second only to the Thompson Seedless in tonnage. It is second to all other grape varieties in flavor and texture. Although the Emperor is attractive in color and grows in large, graceful bunches, it is our least flavorful grape. It has a very low sugar content and a very tough, chewy skin. Its greatest asset, besides its tremendous yield per acre and low price per pound, is its hardiness. It is the darling of the self-service markets because it can withstand rough handling. They arrive in market in August and wind up in March, and they are also imported from Chile during the winter months.

Melons

CANTALOUPES
·
CASABAS AND OTHERS
·
CRENSHAWS
·
HONEYDEWS
·
WATERMELONS

Botanically melons are in the same family as gourds and are related to cucumbers and squash, which also grow on vines. They have a very high moisture content and should have a high sugar content. Even though they are sweet, they have a fairly low calorie count.

All melons (except watermelon, which is a different branch of the same family) are structured similar to winter squash, with a hollow cavity that contains the seeds. Their flesh is seed-free. The watermelon is structured like a cucumber or a summer squash: The seeds are dispersed throughout the flesh, rather than being concentrated in a seed cavity.

Melons come in assorted sizes, shapes, skin colors, and flesh colors. Even though there are hundreds of kinds of melons, only about a dozen varieties are grown commercially, and three of these comprise about 90 percent of the total melon tonnage. The rest of the melon varieties are not as a rule as flavorful as the top three. These three, in order of tonnage and merit, are the cantaloupe, honeydew, and crenshaw. The also-rans, in alphabetical order, include the Casaba, Galia, Green Tendral (Spanish Melon), Juan Canary, Persian, Santa Claus, and Sharlyn. The also-rans should only be considered when the top three are not available.

Choosing a perfect melon requires some degree of expertise, which is rewarded with perfect flavor. Lack of expertise usually results in a flavorless melon.

CANTALOUPES

The cantaloupe, a golden-fleshed melon, is the most reliable and least costly melon on the fruit stand when in season. While it takes some know-how to identify top quality in most of the other varieties, selecting a fine cantaloupe is comparatively easy. All you have to do is select high color (yellow), unbruised, firm melons and allow them a few days to ripen at room temperature. You'll know when they are ripe because they'll give slightly to gentle pressure and will have increased fragrance. Of course, if you settle for soft, soupy, banged-up, moldy melons or those that are grass-green, all bets are off.

Cantaloupes are available year-round but are at peak of season from June through November. In December, January, and February they are imported from Central America and some of the islands in the Caribbean, but these imports are usually overpriced and underflavored. In this three-month period, when all the other varieties of melons are similarly poor and costly, your best bet is to use the far less costly and less risky grapefruit in place of melon.

In February, March, April, and May Mexican cantaloupes are available. These are of better quality but are not always reliable. They too are usually overpriced.

In June the California cantaloupes come on line, and they don't bow out until December. California not only grows the world's finest cantaloupes, usually supplies are more than ample to satisfy the demand and prices are modest. When California cantaloupes are in season, skip the grapefruit. In that time span the grapefruit are below par and high in price. Cantaloupes have a low calorie count and are popular with dieters.

Cantaloupes are grown in many areas of the United States. The farther west, the better the melon. Arizona also grows fine melons. Texas cantaloupes used to be mediocre, but in recent years they have become quite good. As a rule, the cantaloupes grown in the rest of Dixie are no bargain. Most grown east of the Mississippi are long on size and short on flavor. Nearly all are sold locally. There is one exception worth noting: There is a grower named Hand who produces melons in northern New York State. Hand melons are famed for their flavor and texture. However, the demand usually exceeds the supply. They seldom reach the wholesale markets. Hand usually sells his entire crop during the Saratoga racing season.

☐ WHEN TO BUY: *At peak June through November*
☐ WHAT TO LOOK FOR: *Firm, unbruised fruit with light-yellowish skin*
☐ HOW TO STORE: *Ripen at room temperature then refrigerate*

CASABAS AND OTHERS

The following melons at best are not quite as flavorful as a ripe cantaloupe, honeydew, or crenshaw. Consider purchasing these melons only if the top three varieties are not in market.

Casaba 🍃

The casaba, a white-fleshed melon, is my candidate for our poorest flavored melon. When fully ripe, it is juicy but has a squashlike flavor. More often than not, they just aren't ripe. They arrive in August and are available until the end of December. They have a light green skin color when not quite mature and this color changes to a pale yellow as the melon ripens. They can be identified by their flatness at the blossom end and the scored wrinkles in the skin at the stem end. Casabas will gain color and soften if left at room temperature, but they will not get any sweeter.

Galia 🍃

The Galia is fairly new in North America. They are grown in Israel and flown in via jet in November, December, January and February, when most other varieties are out of season. They look like pale-skinned, fairly smooth cantaloupes, but they have a light green flesh. The flavor is quite sweet and they have a nice, smooth, juicy texture. They are usually quite costly. In recent years Galias have also been grown in Latin America and Puerto Rico, but these aren't quite as good as those grown in Israel.

Green Tendral (Spanish Melon) 🍃

The Green Tendral, also known as the Elche Honeydew or the Spanish melon, is shaped somewhat like a football and has a dark green, ridged skin and a cream-colored flesh. It has a very hard shell even when ripe. The best ones are grown in Spain and usually arrive in December. They are available for two to three months at a time when there are few melons in the market and they are usually quite costly.

Green Tendrals are also grown on a limited scale in California and Chile, but neither area produces melons that are as flavorful as the ones from Spain, which, when ripe, can be very tasty. However, identifying one that is ripe is most difficult. If it has the same slight give at the blossom end that identifies a ripe honeydew or a crenshaw, the Spanish melon is probably overripe and will eat mushy. Your best bet is to buy one that is firm at the blossom end and hope for the best.

Juan Canary &

The Juan Canary looks just like a lemon-colored honeydew. It has a white flesh and at best is sweet and juicy. But it is difficult to find one that is at its best. They are either too firm or too soft and seem to go from one extreme to the other in a matter of hours. Since they are in season from July to November, the same period that the top three varieties are at peak of season, skip the Juan Canaries.

Persian Melon &

The Persian melon is superb when it is vine-ripened. It is a large round melon that has a deep orange flesh and a sweet, musky flavor. The skin is heavily netted, and when vine-ripe, the color peeking through the netting is orange. However, in nearly all the Persians that come to market, that color peeking through the webbing is green, signifying that the melons were picked prior to reaching maturity. While vine-ripened Persians are available in California, in New York Hunts Point Market, the world's largest wholesale produce terminal, ripe Persians are extremely rare. Persians are usually available in August and September.

Sharlyn &

This is a fairly new variety and is akin to the Persian melon. It looks like a Persian with white flesh. Supplies are very limited, and it's too early to pass judgment on them. However, most of the ones that have arrived at market have been ripe and have had fine flavor.

Santa Claus Melon 🍂

The Santa Claus melon got its name because at one time, prior to our importing melons from Spain, Israel, and Chile, it was the only melon that was in market during the Christmas season. It is large, dark and light green striped, and shaped like an oversized football. The texture and the color of the flesh is similar to that of a honeydew, but it is firmer and not as sweet. It isn't as bad as a casaba or a Juan Canary, but it's not in the same league as the honeydew, cantaloupe, or crenshaw.

CRENSHAWS

The crenshaw is a fine, golden-fleshed melon. It is large in size and very thick-meated because it has a small seed cavity. When vine-ripened it is sweet and juicy, but it will taste a bit like squash if it isn't fully matured. The peak of season is from August to December.

There are two types of crenshaw: The original crenshaw is a golden yellow in skin color; the hybrid, a pale yellow (almost white). The golden-skinned crenshaws are the better melons.

Unlike the honeydew, which must be ripe at the time of purchase, a golden crenshaw that is firm will ripen up at room temperature. However, one that has a dark green skin will never attain full flavor and ripeness.

Crenshaws are quite huge, often weighing more than ten pounds. A melon that large is fine for a restaurant but more than the average-sized family can handle. Very often when the crenshaws are running very large in size some markets will cut them in half and display them with the cut side up. Even a green-skinned crenshaw looks pretty good after it has been cut. Check the skin color and texture prior to purchasing a half. If it isn't golden in color and doesn't feel velvety, it isn't fully ripe. While a partially ripe uncut melon will ripen at room temperature, a cut piece of the same melon when stored in the refrigerator will ripen no further.

☐ WHEN TO BUY: *At peak August, September, and October*
☐ WHAT TO LOOK FOR: *Firm, unbruised fruit with yellow skins*
☐ HOW TO STORE: *Ripen at room temperature, then refrigerate*

HONEYDEWS

By far the sweetest melon, if not the sweetest fruit, that grows is a vine-ripened honeydew. Unfortunately, they are not easy to come by. I've seen huge displays of honeydews in supermarkets, with not one vine-ripened honeydew among the hundreds being offered. This scarcity of ripe honeydews is especially prevalent in markets that are self-service. Along with vine-ripened tomatoes, vine-ripened honeydews are too delicate and fragile to withstand rough handling by overzealous shoppers in self-service markets. Therefore, these markets usually buy immature, rock-hard melons to cut down on losses due to less than gentle handling by consumers.

All honeydews, whether ripe or immature, are light green in flesh color, except for a recently developed orange-fleshed variety that has all the prospects of a real winner.

How does one identify the elusive vine-ripened honeydew? There are four criteria: fragrance, skin color, skin texture, and the slight give to gentle pressure at the blossom end.

A ripe honeydew is as fragrant as a flower. The blossom end, as opposed to the stem end, will be more fragrant. A hard, immature honeydew won't have any odor.

The skin color of a vine-ripened honeydew is the color of creamery butter. If it is chalky white, pale green, or canary yellow, it is not a good sign. However, if the skin has a netting or a ribbing and especially if the skin is blemished with freckles, don't pass up the melon. These freckles are clues to high sugar content.

The feel and texture of the skin are foolproof keys to quality. The skin of a truly vine-ripened honeydew will feel velvety and slightly tacky—like a freshly powdered baby's bottom. The unripe honeydew will have a skin texture that is smooth and slick.

The blossom end of the honeydew should have a slight yield to *gentle* pressure. The word *gentle* is emphasized because too many fine melons are destroyed by well-meaning but overzealous shoppers. Don't apply any more pressure than you would to a grape.

When you purchase a honeydew, what you see is what you get. Unlike a firm, high-colored cantaloupe or a firm, golden-colored crenshaw, a firm honeydew will not ripen after purchase. If you settle for a hard honeydew, no matter if you incubate it for months on end, it will not ripen or get any sweeter.

Sometimes an unscrupulous produce manager who is up to his ears in rock-hard honeydews will bang them on the blossom end. In the parlance of the produce trade, this practice is known as hammer-ripening. Don't fall for this skulduggery. If a honeydew is soft at the blossom end but the rest of it feels slick and hard as a rock, it probably was hammer-ripened.

Having cautioned you against purchasing very firm honeydews, let me warn you about those that are too soft. A soft, soupy honeydew, especially if the seeds go "slurp" when you shake it gently, is probably overripe. In an overripe honeydew the flesh will show shatter (be broken up) or will appear to be frozen and look glossy. As a rule, when you cut open an overripe honeydew, you had better have a towel or even a mop handy to catch the liquid.

An overripe honeydew may be as sweet as sugar even while being at the point of fermentation. Eating it could result in a very upset stomach (melon colic?). Very often a superb honeydew is too sweet and will cause discomfort. If you have one that is too sweet, sprinkling it with a few drops of fresh lime or lemon juice will solve the problem and could even enhance the flavor of the melon.

California grows the finest honeydews. Those grown in Arizona, Texas, and Mexico aren't quite as good. Although the California honeydews make their annual debut in June, the peak of season is August, September, and October. In that time period we get honeydews that are grown in the Turlock area of central California. These are the finest in the world. There are four major growers in that area and they identify their products by attaching gummed labels to each melon. The trade names to watch for are: King of the West, Peacock, Sycamore, White House, and White Star (the last two belong to the same grower). While it is possible to get a too firm Turlock that was picked prior to reaching full maturity, the above labels have a great track record.

In one sentence, the way to select a perfect honeydew is to choose one that smells like a flower, is the color of butter, has a slight give at the blossom end, and feels like a baby's bottom. If you can't find a truly ripe honeydew, switch to the more reliable cantaloupe or check out the crenshaws.

☐ WHEN TO BUY: *At peak July through October*
☐ WHAT TO LOOK FOR: *Unbruised, fragrant, cream-colored fruit that is velvety but not soft*
☐ HOW TO STORE: *Refrigerate immediately*

WATERMELONS

Watermelons originated in Africa at least four thousand years ago. They are edible gourds and are botanically in the same family as squash, cucumber, and cantaloupe.

In the United States, Florida, Texas, and California are the top producing areas, and most of the other southern states have substantial crops. They are even grown as far north as Delaware. Domestically grown watermelon are available from April to November. When our watermelons are out of season, we import them from Mexico and Central America.

There are many varieties of watermelon and they come in assorted sizes, shapes, and skin colors. The flesh can either be red or yellow, but 99 percent of the watermelon sold in the United States are of the red-fleshed varieties. They range in size from about as small as a honeydew (these are called Ice Box watermelons) to exceeding fifty pounds. Some watermelons grown to compete in contests reach weights of two hundred pounds. They are either oval or round. In skin color, some are a solid dark green, some are a solid pale green. Many varieties are variegated with alternating dark and pale green stripes. The size, shape, and skin color have no bearing on the quality, flavor, and ripeness of a watermelon.

The top-selling varieties include the pale green, oval Charleston Grays, the dark green, round Black Diamonds and Peacocks; and the Sugar Sweets and Klondikes, which are striped, oval melons. The Sugar Babies are small, round, dark green Ice Box melons. Equally good flavor and texture can be found in all varieties.

Seedless watermelons have been produced by experiments in the agricultural schools, but attempts to grow them on a commercial scale have not yet been fruitful. The seedless watermelons that I have sampled were not nearly as sweet as watermelons with seeds. The few yellow-fleshed watermelons I have tried have also lacked flavor. There is a common but inaccurate belief that the blacker the seeds, the more mature and flavorful the watermelon. However, the color of the seed is determined by the variety, not the maturity, of the melon. Some of our most flavorful varieties have light-colored seeds.

Prior to cutting open the watermelon, it is very difficult, if not impossible, to judge its stage of maturity. Professional produce buyers will never buy a load of watermelon without cutting several random samples. These melons do ripen after being severed from the vine. The length of time that has elapsed since harvesting can be determined by the condition of the pigtail stem. When you purchase a watermelon, if the stem is fresh and green, the melon is probably too immature to cut and needs a few more days of ripening. After a few days the stem will shrink and discolor but will still be attached to the watermelon. This is a sign that the watermelon has reached the desired maturity. After another few days, the pigtail stem will part from the watermelon. A tailless watermelon may be overripe.

Thumping a watermelon to check its ripeness is an exercise in futility. The only foolproof way to judge the ripeness and the texture of a watermelon is to cut it open. Retail markets sell cut watermelons by quarters and halves. Purchasing a whole uncut watermelon is like buying a pig in a poke. Always buy a quarter or a half and check it out for color. Even if a whole watermelon is priced lower than two halves, it is wiser and safer to buy the cut melon.

The perfect watermelon will have a firm, dark red flesh. If the flesh is pale and pink, it isn't quite ripe and will lack sweetness. If the flesh is soft or shattered, the melon is overripe and will have a poor flavor and texture. Even in a perfect watermelon, the blossom-end half is always slightly riper and sweeter than the stem-end half.

Even though watermelons are now available year-round, they seem to taste better when the mercury is in the eighties or nineties. Unlike other melons and most fruits, which lose some flavor when served chilled, ice-cold watermelon is at its flavor best.

☐ WHEN TO BUY: *At peak June through August*
☐ WHAT TO LOOK FOR: *Never purchase a whole, uncut watermelon if it can be avoided; flesh should be dark red and firm*
☐ HOW TO STORE: *Refrigerate cut fruit immediately; store uncut fruit at room temperature*

Nuts

ALMONDS
·
BRAZIL NUTS
·
CASHEWS
·
CHESTNUTS
·
COCONUTS
·
HAZELNUTS (FILBERTS)
·
MACADAMIA NUTS
·
PEANUTS
·
PECANS
·
PISTACHIOS
·
ENGLISH WALNUTS

T he nut is a dry fruit consisting of a shell-encased kernel or seed. Except for the peanut, which is not a true nut but a legume, nuts grow on trees.

ALMONDS

Although almonds are native to the Mediterranean basin, California is the world's number one producer and exports them to many nations. They are related to the stone fruits, such as apricots and peaches. If you crack the pit of an apricot or a peach, the inner seed will resemble a shelled almond, and the almond is indeed the stone or pit of the fruit.

There are two types of almonds: The bitter almond, which is inedible, is used to make flavorings and extracts; the sweet almond is the kind you buy in a food store. Whether in its shell or shelled, sweet almonds are used as table nuts. They come in two types: The ones with thin shells are called paper-shelled almonds; the ones with the thicker, harder to break shells are less costly but usually yield a better flavored nut.

BRAZIL NUTS

Brazil nuts grow on tall trees that bear nuts that look like coconuts. But unlike the coconut, which is hollow and filled with liquid, the "coconut" of these Brazilian trees contains fifteen to twenty closely packed Brazil nuts. The Brazil nut is three-cornered and the segments fit together much like the segments of citrus fruit.

The Brazil nut has a very hard shell and a very hard nutmeat. But even though they are of a hard texture, they are quite perishable. If they are not stored in cool dry areas, they often get rancid. Twenty or thirty years ago we had a far higher percentage of rancid Brazil nuts, but today this problem is much less common because of improved methods of storage and more rapid transit.

CASHEWS

The cashew tree is native to South America, but India is by far the world's largest producer of this nut. The cashew is of the same genus as the mango. But unlike the mango, where the flesh of the fruit is used and the stone is discarded, with the cashew it is the stone that is edible and the flesh has no value.

Cashew nuts are never shipped to market in the shell. These nutmeats are roasted and salted prior to being offered for sale. In recent years, due to diets that discourage salt intake, unsalted cashews are available and their sales show a steady annual increase.

CHESTNUTS

There used to be three kinds of chestnuts: European, Oriental, and American. The American chestnuts were the most flavorful, but unless you are a very senior citizen, you have yet to, and probably never will, sample one. Millions of America's chestnut trees were wiped out in the early 1900s by an incurable disease called the chestnut blight. The European, and especially the Oriental, strains are resistant to the blight that totaled the American trees.

Today, most of our chestnuts are imported each fall from Italy, although some are also brought in from Portugal and the Far East. They are graded by size—the larger the nut, the higher the price. They used to be graded AAA, AA, and A. Now we use a metric measure: 44–46, 48–50, and 60–65. The 44–46, which corresponds to the old AAA, means that there are 44–46 nuts to a kilo (2.2 pounds).

The chestnut tree is a large shapely tree that is also prized for its timber. It blossoms and then produces a round green prickly burr that contains two or three closely fitting chestnuts.

These nuts may be eaten raw, but they are usually roasted. Sometimes they are boiled in a sugar syrup. Until recently, before prices sky-rocketed, it was common to see vendors selling hot roasted chestnuts in the streets of our larger cities. The fine warmth and flavor of chestnuts roasting in charcoal braziers are second only to the fragrance.

In France a common dessert is *marrons glacé*—shelled chestnut meats that have been boiled in sugar syrup with vanilla or liqueur flavorings. In the United States you can buy these bottled chestnuts in the more posh retail food stores.

COCONUTS

Coconuts grow on palm trees and are related to dates. Just as the date is the staff of life for people who dwell in the desert, the coconut plays a similar role for those who live in the tropics.

The coconut, the world's largest nut, has a double shell. The outer shell is green in color, fibrous in texture, and shaped like an oversized football. It is most difficult to remove unless you have a machete. Natives of areas that produce coconuts whack off the outer shells in no time. Fortunately for us, coconuts arrive in our retail markets without it.

The inner shell is round in shape, brown in color, and covered with hairy brown fibers. It is as hard as a rock and very brittle. Removing this inner shell is no easy task either. At one end of the coconut you'll find three smooth, penny-sized depressions, called eyes. Puncture a hole in two of these eyes—the best instrument to use is a tenpenny nail. Drain out the liquid, which is called milk and can be used as a beverage.

After you have milked the coconut, you have three options and each requires a hammer. The first is to start hitting the coconut on all sides until you crack the shell, then pry it off. Or you can put the coconut in a 300° oven for about a half hour, or put it in your freezer for a few hours. Even with these last two methods, the heated or cooled shell will still pry away from the coconut meat more easily if you hit it with a hammer. The shell that has been chilled will be more brittle and will shatter more easily.

Is it any wonder that most people feel that the end product isn't worth the effort, and, if they do use coconut, buy it in a tin? However, if you think fresh coconut *is* worth the effort, here's how to select a good one: Purchase one that is heavy and has a lot of liquid. This can be determined by shaking the coconut. If you hear a sloshing sound, you have a winner. If you are greeted with silence, don't buy it. Most adults won't bother with a fresh coconut, but the small fry have a ball. Coconuts are available year-round, but are more plentiful during the fall and winter months.

HAZELNUTS (FILBERTS)

Hazelnuts are also known as filberts because in European folklore they are ready for harvest on August 22—St. Philbert's Day.

Depending on the variety, hazelnuts grow on both trees and shrubs. The ones that grow on trees are of superior quality. By far the largest producer is Turkey, with Italy a distant second. In the United States, Oregon produces a small crop.

Hazelnuts have a smooth, hard shell and look like round brown marbles. Some varieties are slightly elongated in shape. Hazelnuts are used as a table nut or are added to shelled-nut mixtures sold in tins. They are highly prized by bakers and candy makers.

MACADAMIA NUTS

The Macadamia nut originated in Australia and was transplanted to Hawaii, where it is now a major cash crop. In recent years some Macadamia trees have been transplanted to California, but that crop is still too small to play a significant role in the marketplace.

The flavorful, crunchy Macadamia is encased in an extremely hard shell and is always sold to the consumer already shelled by a special machine. It is the Rolls-Royce of the nut world. They are very costly, but well worth the price. The Macadamia, along with the pineapple and the hula skirt, has become a symbol of Hawaii. En route from the mainland in an airplane, flight attendants pass out Macadamia nuts to herald the approach to the islands.

PEANUTS

Peanuts are native to North America. They were cultivated by the Indians prior to the arrival of the European explorers. The peanut is not a tree nut, but rather a member of the legume family, which includes peas and beans. But unlike other legumes, the peanut is produced beneath the surface of the earth, which is why it is called a ground nut in some areas.

Today, peanuts are an important food staple to millions of people in the underdeveloped areas of Africa and Asia. In the United States it is an important cash crop south of the Mason-Dixon Line. Georgia is no longer the number one peach state but is still the number one peanut state.

Raw peanuts, when processed, yield a fine, clear vegetable oil. When roasted, peanuts are used as snacks, and, more important, they are ground into peanut butter. This supplies our small fry with half the ingredients of their favorite sandwich.

Peanuts are by far the least costly, most plentiful, nut. The shell is thin and can be easily cracked by finger pressure.

Peanuts in the shell are available year-round, but many people prefer to buy them packaged in tins and jars.

PECANS

The pecan is native to North America. Texas and Georgia produce nearly all the commercially grown pecans. In North America it is second only to the walnut in popularity. It is sold as a table nut and is also used by bakers and candy makers. Pecan pie is one of America's most traditional desserts.

Pecans have flavorful nutmeats encased in elongated, smooth shells. The rather drab-colored shells are often colored and polished to enhance their appearance. The thinner-shelled varieties command higher prices than those that are hard-shelled. However, the thickness of the shell has no bearing on the quality of the nutmeat.

Pecans in the shell are available year-round, but in the summer months they aren't nearly as good as the ones available in the fall and winter. Shelled pecans are also available packed in tins.

PISTACHIOS

Pistachios are related to cashews and mangoes. Iran and Turkey are the world's largest producers of this flavorful nut, but California also produces a small but annually increasing tonnage.

The pistachio has a small hard shell that contains a fine-textured nutmeat. Its light green color is particularly unusual because most, if not all, other nutmeats are white in color. The shell of the pistachio is almost colorless but is often dyed with a harmless red coloring. While this may add to the appearance and the cost of the pistachio, it plays no role in its flavor or texture.

Even when using a nutcracker, the pistachio's small hard shell is very difficult to crack, which is why they are partially cracked prior to shipment to the marketplace.

Pistachios are available twelve months of the year.

ENGLISH WALNUTS

Although we know them as English walnuts, they originated in the Middle East, in areas near the Caspian Sea. Few, if any, are grown commercially in Britain.

The United States is now the world's largest producer of walnuts, and by far the lion's share of our crop is grown in California, where the walnut groves are a tourist attraction. As seen from the road, the trees, almost identical to each other in size and shape and spaced equidistantly apart, look like soldiers standing at attention.

The California crop is harvested early in the fall in order to reach the marketplace in time to meet the Thanksgiving demand. However, there is usually a carryover from the previous year's crop that is stored under refrigeration during the summer months. Even though walnuts and many other nuts are stored under ideal conditions, it is usually a good idea to buy them shelled in a tin prior to and at Thanksgiving. By Christmas, the next big holiday for walnuts, the old crop of nuts will probably have been used up and the nuts you purchase will surely be of the new crop.

Black walnuts, which have a good flavor but are encased in almost unbreakable shells are usually not sold in food markets, although some shelled black walnuts are sold to bakeries.

Vegetables and Herbs

The Cabbage Family

BROCCOLI

·

BRUSSELS SPROUTS

·

CABBAGE

·

CAULIFLOWER

·

KOHLRABI

C abbage, which originated in Europe, comes in assorted sizes and shapes and in two colors, green and red. The various members of the cabbage family include broccoli, Brussels sprouts, cauliflower, collards, kale, kohlrabi, and Oriental cabbage (Chinese and Nappa) as well as the more common head cabbage. All varieties are nutritious, high in fiber, easy to grow, and hardy, and all thrive in cool, almost freezing, weather. All types of cabbage are at peak of season in the cooler months of the year. All are similar in flavor and have a similar odor when cooking and most are excellent when served raw as well as when cooked. If we added the tonnage of all the different types of cabbage, along with that of turnips, we would arrive at a total that might exceed any other fresh fruit or vegetable. Turnips are closely related to cabbage, but since they are grown below the surface of the ground, they are discussed under the category of root vegetables. Although collards and kale are listed above, since they don't head and are prepared like spinach, they are discussed along with leafy greens.

All members of the cabbage family are available twelve months of the year and should be refrigerated immediately after they are purchased.

BROCCOLI

Broccoli, the Italian branch of the cabbage family, was not widely known or used in the United States until about twenty-five years ago. In 1920 Stephen and Andrew D'Arrigo planted trial fields of broccoli in California using seeds imported from Italy. Today, America's largest grower of broccoli is D'Arrigo Bros. Co., which farms more than five thousand acres near Salinas. The surrounding area from Santa Barbara to just below San Francisco produces almost 90 percent of the nation's annual production. Sixty years ago fresh broccoli could be purchased only in Italian neighborhoods. Today it is one of our best-selling fresh, as well as frozen, vegetables.

While California is by far the number one producer of fresh broccoli, substantial crops are produced in Texas, Oregon, and Arizona. Fresh broccoli is available twelve months of the year, with the peak of season from October through May. The light part of the season is usually July and August, but supplies are usually adequate to meet the demand.

When shopping for broccoli, the key words are *firm* and *green*. The undesirable words are *limp* and *yellow*. Top quality broccoli has firm, compact clusters of dark green buds. In some varieties the buds have a slightly purplish hue, which is a mark of quality. Never purchase broccoli that has started to yellow and especially avoid broccoli that has buds that have opened up and show tiny yellow flowers.

Broccoli is fairly perishable and when it isn't fresh it loses flavor. Don't purchase broccoli more than a day or two prior to use, and in the interim wrap it with plastic film and store it in the vegetable bin of your refrigerator. Broccoli is not only an excellent cooked vegetable, it may be served raw in salads or with dips.

BRUSSELS SPROUTS

Brussels sprouts are the newest members of the cabbage family, having been around for only a few hundred years. They grow on a very unusual yet attractive plant that from a distance looks a bit like a miniature green papaya tree. The leaves are on the top of the plant and the tiny heads (called sprouts) completely surround the stalk. Brussels sprouts look like miniature heads of green cabbage.

Brussels sprouts thrive in cool, damp weather and for some reason are at their best when grown not too far from the ocean. California is by far the number one source, but we do import a fair amount from Mexico during the winter months. During the fall until the first fairly heavy frost, Long Island in New York has a large, top quality crop.

The peak of the California season is from October through March; there is some slackening of supplies during the summer months. However, if fresh ones aren't available or are overpriced, frozen Brussels sprouts are one of our better frozen vegetables.

Most fresh Brussels sprouts are marketed in sixteen-ounce, film-covered, waxed-paper cups. As with fresh broccoli, it's go on green and avoid yellow. The smaller, firmer, and greener the sprout, the better the flavor. Soft, flabby ones, even if green, are less desirable than the hard, compact sprouts. Those with yellow leaves are undesirable.

Fresh sprouts aren't very perishable and should last for at least a week if you store them in the coolest area of your refrigerator.

CABBAGE

While there are countless varieties of head cabbage, they can be divided into three groups: green, red, and Savoy (curly). Nearly all cabbage falls into the first group.

Green cabbage is available twelve months of the year. Barring unusual weather conditions, supplies are always ample and prices are always reasonable. Along with carrots, green cabbage is almost always the least costly, yet one of the most nutritious, fresh vegetables at the poduce counter. It is grown in every state, the top producers being Florida, Texas, California, New Jersey, and New York. Most of our sauerkraut is canned in upstate New York.

Cabbage isn't one of our most elite vegetables because it gives off a strong odor during the cooking process. There are many old wives' tales about checking the odor of cooking cabbage. One is to drop a walnut (shell and all) into the boiling water. However, the best bet is to open a couple of windows or, if you have one, turn on the exhaust fan. Cabbage is not only used as a cooked vegetable but is the main ingredient in making cole slaw.

Green cabbage is also known as new cabbage because it arrives at market with all of its green outer leaves. The fresher the cabbage, the greener the leaves. New cabbage is available twelve months of the year. In Grandma's era, during the winter and spring in the northern half of the United States, new cabbage was out of season. However, since cabbage keeps for months without spoiling, Grandma used cabbage that had been stored in barns. This was called old cabbage, was white rather than green in color, and was usually as hard as a rock. Old cabbage is still used by the makers of the commercial cole slaw sold in supermarkets. You can recognize it because it is white rather than pale green.

Red cabbage, except for its color, is identical to the green type, but usually sells at a higher price. Savoy cabbage has lacy, curly leaves. The head is never quite as hard as that of green or red cabbage, but it has a more delicate flavor and texture. Savoy cabbage usually sells for about the same price as red cabbage.

When purchasing cabbage, select solid, heavy, fresh-looking heads. Avoid those that have flabby yellow leaves.

CAULIFLOWER

Cauliflower is the most genteel member of the cabbage family. It has a milder flavor, isn't quite as "fragrant," and usually commands a higher price than its more common cousins. Like all other cabbages, it is at its flavor best and in most ample supply during the cooler months although available twelve months of the year. While there are fair-sized crops grown in Oregon, Texas, New York, and Florida, by far our largest supplier is California. Supplies of California cauliflower exceed the total output of the rest of the states plus Canada.

While there are some green-and purple-headed types, 99 percent of the cauliflower grown is of the white or ivory-headed varieties. Cauliflower has always been widely used as a cooked vegetable, but in recent years it has gained much favor when used raw in salads and with dips.

Cauliflower grows on a leafy green plant. The inner leaves (called jackets because they cover the head) look and taste like collard greens. Fresh cauliflower used to come to market with these green jackets, but today nearly all of it arrives minus the greens and covered with plastic film.

When selecting cauliflower at the marketplace, choose heads that are heavy, compact, and free from discoloration (brown spots). Avoid heads that aren't solid and are starting to spread apart, which is a sign of overmaturity. Occasionally cauliflower will have a slightly granular appearance, a condition called ricey. While this won't affect the flavor, it is slightly less desirable because ricey cauliflower is light in weight and won't yield as many servings as a similar-sized solid head.

KOHLRABI

The kohlrabi is an unusual-looking cabbage. It resembles a root vegetable rather than head cabbage and is sold banded in bunches. A bunch of kohlrabies looks somewhat like a pale green bunch of beets. Yet unlike the beet, the bulb of the kohlrabi is produced aboveground.

While it is still fairly popular in Germany and middle Europe, the kohlrabi has few fans in the United States. Each year less and less kohlrabi arrives at the marketplace, yet it deserves more popularity. Kohlrabi has a nice mild flavor when cooked and is surprisingly crisp, crunchy, and juicy when served raw. Cut off the tops if you plan to store the kohlrabi for a few days. If the tops are green and fresh, they can be cooked like spinach or any other leafy green. Cook as you would cook cabbage (boil or steam). Store in the refrigerator.

It is at peak of season from late spring to early fall. However, since it is also grown on a limited scale during the winter months in Florida, Texas, and California, kohlrabi is available twelve months of the year. Along with the more common pale green variety, there is also one that is a very pretty purple-red in color.

Select kohlrabies that look fresh and crisp. Avoid those with wilted yellow leaves. Size is important, because if the bulbs are too large, they may be woody. For size, select kohlrabies that are no bigger than the acceptable size of beets.

Leafy Greens
(for Cooking)

SWISS CHARD

·

COLLARDS

·

BEET GREENS

·

MUSTARD GREENS

·

TURNIP GREENS

·

KALE

·

SPINACH

T hese leafy greens are used as cooked vegetables. While they belong to diverse botanical families, they have in common a leafy structure and similar methods of preparation. Except for spinach, which plays a dual role and may be used as a raw salad green, the leafy greens listed below are used as cooked vegetables.

All leafy greens are available twelve months of the year and should be refrigerated immediately after they are purchased.

SWISS CHARD

Swiss chard is related to fresh beets but doesn't form a bulbous root. It is usually available twelve months of the year but is fairly perishable during the warmer months. There is an Oriental variety of Swiss chard that is called bok choy. While it has a different name and a slightly different appearance, it is almost identical in flavor and texture to the domestic chard.

Swiss chard is usually marketed in bunches. It has a green leaf with fairly thin white ribs. There is also a very attractive red-leafed variety that is seldom available in retail stores. However, many home gardeners prefer it because it is more colorful yet equally flavorful. If you have a home garden, you'll find that Swiss chard will have a greater yield and require far less care than spinach, which often burns up in hot weather.

Swiss chard is superior to spinach because when spinach is even slightly overcooked it dissolves to a gooy, almost slimy, texture and the pretty green color fades to a dull olive drab. Swiss chard doesn't break down; it remains green and doesn't get slimy.

Select Swiss chard that has crisp green leaves and avoid those that are limp, wilted, or have started to turn yellow.

COLLARDS

Collards are also known as collard greens. They look like large, flat green cabbage leaves and taste almost exactly like green cabbage but with a slightly smoother texture.

Collards are primarily grown and sold in the rural areas of the southern states and in the large industrial cities of the Midwest and Northeast, where there are large groups of people who have migrated from the South.

The traditional Southern recipe for using collards is to cook them along with smoked ham hocks or salt pork, a dish that is traditionally referred to as "A Mess of Greens." This may explain why the collards seldom appear on restaurant menus up North.

Collards keep well for about a week or so if stored in a refrigerator, but it is advisable to buy them as you need them to ensure top flavor and nutritional value. Select young, green, velvety-feeling, crisp leaves. Avoid those that have started to discolor (by turning yellow), look limp and wilted, or show traces of insect damage (holes in the leaves). Collards are available twelve months of the year but, like most other members of the cabbage family, are at their flavor best when they have been exposed to cool weather.

BEET GREENS

Beet greens are the tops of young garden beets. Occasionally when you purchase a bunch of young, fresh, locally grown beets, the greens will be fresh and tender enough to use as a cooked vegetable.

Young, fresh beet tops with tiny beets attached arrive in the market early in the spring, brought in by farmers who have thinned out their beet crop. If the tops are fresh and crisp, they are a real taste treat. Hothouse beet tops, grown in greenhouses, used to be available during the winter months, but of late they seem to have disappeared from the marketplace.

If you have a garden, pull some beets while the bulbs are not much bigger than marbles. You will find that these greens won't break down when you cook them and that they have a flavor, color, and texture superior to spinach.

MUSTARD GREENS

Mustard greens look like a more delicate version of kale and are a lighter green in color. These greens are a big item in the South, but don't sell too well in the rest of the country.

Shop for mustard greens as you would for kale. The leaves may be dark or light green in color or even have a trace of bronze. Avoid those that are limp and yellow.

TURNIP GREENS

The green tops of white turnips are marketed as turnip greens or turnip tops. Like collards, they are primarily grown in the South, where they are a staple in the diets of people in lower income groups.

Occasionally the turnip greens are sold in bunches, with the small white turnips attached. However, they are usually marketed without the turnips.

When shopping for turnip greens, select those that have fresh green leaves and avoid those that are limp or yellow. Check the leaves for insect damage. If there are a lot of small holes, the bugs have already had a party.

The green tops of the yellow turnip are called Hanover greens. They are slightly less desirable than the greens produced by the white turnip.

Where there is demand for this product, turnip greens are available year-round. Prepare them as you would spinach.

KALE

Kale is a curly-leaf member of the cabbage family. Like collard greens, it is available year-round but is at its flavor best when exposed to cold weather. It has the traditional cabbage flavor and aroma when cooked.

Choose kale that has crisp, colorful leaves. The color may be dark green or slightly blue. Avoid wilted, limp, yellowish-colored leaves. Ornamental kale, offered by florists during the cooler months, has a pink, purple, or white heart and is used for decorations, but it is too costly to use as a vegetable.

SPINACH

Fresh spinach is available twelve months of the year but is most plentiful in the spring. Since it is fairly perishable and breaks down in very hot weather, it is usually of better quality in the cooler months. Although spinach is grown locally in most of North America, Texas is the number one producer, and Colorado, New York, New Jersey, and Ohio also have major crops.

There are many varieties of spinach, but they can be broken down into three groups: the Savoy, which has crinkly leaves; the semi-Savoy, which has leaves that aren't as crinkly; and the flat leaf. All three types are crisp and dark green when fresh and limp and yellow when aged.

The flat-leaf spinach is grown mainly in California. This type is very popular on the West Coast, but it is fairly new in the East. At first this spinach sold poorly among easterners, but its sales there are gaining momentum. Flat-leaf spinach usually comes to market tied into bundles. The Savoy and the semi-Savoy types are sold loose by the pound.

Spinach is also marketed in ten- or twelve-ounce cello bags. This bagged spinach has been washed and the stems have been clipped. It is a good product if handled properly, and it is usually much more costly than unbagged spinach. However, if it is not handled properly either by the retailer or the wholesaler, it often shows signs of trouble. Especially during the warm weather, if not refrigerated properly it will break down and get slimy and show decay.

Skip fresh spinach if it feels very gritty or sandy, especially if you are purchasing the crinkly Savoy type. The sand is most difficult, or even impossible, to remove. The California flat-leaf spinach does not pose this problem.

Lettuce and Bitter Salad Greens

❧

LETTUCE
·
BIBB (LIMESTONE)
·
BOSTON (BUTTERHEAD)
·
ICEBERG (CRISPHEAD)
·
GREEN LEAF
·
RED LEAF
·
ROMAINE (COS)

BITTER SALAD GREENS
·
ARUGULA
·
BORAGE
·
DANDELION
·
BELGIAN ENDIVE
(WITLOOF CHICORY)
·
CURLY ENDIVE
(CHICORY)
·
ESCAROLE
·
RADICCHIO
·
WATERCRESS

T he salad greens are divided into two groups: those that are mild in flavor, usually known as lettuce, and those that have a sharper, more bitter flavor. Where there is consumer demand, all of the items are available year-round.

As a rule in America, salad greens are served raw, but in Europe they are often used also as cooked vegetables. Spinach, which performs a dual role as a fresh salad green and as a cooked vegetable, is covered on page 143.

All lettuce and bitter salad greens are available twelve months of the year. All are perishable and should not be puchased more than a day or two prior to use. They should be stored under constant refrigeration.

LETTUCE

Bibb lettuce, also called limestone lettuce, has a very small head that has dark green leaves and a soft texture. It is often grown under glass, but does equally well when grown outdoors. This is a highly prized variety that frequently appears on menus of more elite restaurants. It is delicate, sweet, and tender, and has more flavor than most other head lettuces. Bibb lettuce is usually fairly costly in price.

Boston lettuce is also known as butterhead lettuce. While it has a fairly large head, it isn't nearly as solid as iceberg lettuce. The leaves are pale green and are loose, soft, and thin. Since it is easy to separate whole leaves from the head, this variety is often used as a bed or a boat for serving other foods. Boston lettuce usually sells at moderate prices.

Leaf lettuce, which doesn't form a compact head, is also known as salad-bowl lettuce. For years it has been a favorite of home gardeners, but only recently has it been grown commercially on a large scale. Leaf lettuce has been very well received at the produce stands and by the restaurant trade. It has a mild flavor and a fairly crip texture. There are several varieties of leaf lettuce. It is usually green, but a reddish variety is rapidly gaining favor. Leaf lettuce is highly prized in Latin American neighborhoods, where it is called *lechuga*. It usually sells at moderate prices at all produce stands.

Iceberg lettuce, also known as crisphead, is the most common salad green. Nearly all the lettuce sold in the United States is of the iceberg variety. Dieticians, gourmets, and most Europeans downgrade iceberg lettuce, claiming that it lacks flavor, character, and nutritional value, but it is by far the people's choice. In this country, iceberg lettuce ranks a close second to potatoes in tonnage as the most used fresh vegetable.

There is always an ample supply of iceberg lettuce. California is the number one growing area, with Arizona coming in a distant second. It is grown to some extent in most states. Depending on the time of year, prices are in the moderate to modest range. Judging by year-round, and year after year, average prices, iceberg lettuce is one of the best buys at the produce counter. This is a very solid head. The leaves are tightly packed together and difficult to remove. For this reason iceberg lettuce is often served in quarters.

Romaine lettuce, also known as cos lettuce, dates back to the Roman Empire. Unlike the other head lettuces, which are round, the romaine has a long, narrow leaf. This is a very crisp, sweet-flavored variety. The outer leaves are slightly coarse and dark green in color, but romaine has a golden heart. This is the variety used to make a Caesar salad. It usually sells at moderate prices.

The trick to purchasing fine lettuce of any variety is to look for heads that look very fresh, crisp, and colorful. Avoid those that look limp or tired. Especially avoid those that have brown discolorations. All varieties of lettuce are perishable and, if stored, should be well wrapped and placed in the coldest area of your refrigerator. They will be at flavor best and at their crispest texture when served cold. Limp or tired-looking lettuce leaves are usually beyond reviving, so keep the lettuce in the refrigerator until you are ready to serve it.

BITTER SALAD GREENS

Arugula, also called rocket salad in some areas, can best be described as an Italian watercress and looks somewhat like the greens of radishes. It has a sharp, pungent, tangy flavor. Arugula used to be found only in Italian neighborhoods and was available from late spring through early fall. It is now rapidly gaining favor and is available year-round in most of the larger retail markets. Since it is fairly perishable, purchase it only if it looks very fresh. Wrap it well and keep it refrigerated. Use it sparingly because it can easily overpower the other ingredients in the salad. Like most Italian favorites, arugula goes well with fresh tomatoes.

Borage, called sandpaper in the produce trade because of its hairy green leaf, is a seldom used salad item and probably isn't offered for sale in your local market unless you live in an Italian neighborhood. Because it is quite aromatic, it is used to flavor salads and other cooked vegetables. In some parts of Europe borage is brewed as an herbal tea and is used for medicinal purposes.

The dandelion you eat is the same plant that is a nuisance when it shows up on your lawn. In Europe wild dandelion that grows in the fields and along the roads is gathered and used as a salad green. In the United States it is cultivated on a limited scale. It is not a big seller because most people object to its very bitter flavor. It is usually used as a raw salad green but can also be served as a cooked vegetable.

Belgian endive (witloof chicory) is probably our most aristocratic and costly salad green. It used to be available only during the cooler months, but now that it is flown in via jet, it is available year-round. However, it is more costly and not quite as fine in quality during the summer months. Nearly all the Belgian endive sold in the United States is grown in Belgium. An insignificant amount is imported from Chile during the summer months, but very little, if any, Belgian-type endive is grown domestically. During World War II some was grown in Michigan, but the product was below par and was discontinued at the end of the war.

Belgian endive dates back only to the mid-nineteenth century. It was discovered by accident by a Belgian farmer who had stored then forgotten some chicory roots in his barn to use as fodder. Some time later, while gathering the chicory roots to feed his pigs, he noticed that some of them had sprouted crisp white leaves. He tasted the leaves, and what is now a multimillion-dollar industry was thus born.

The word *endive* in Belgian endive is a misnomer. In Europe it is properly called witloof chicory. (*Witloof* in the Flemish tongue means white leaf.)

The Belgian endive is produced by forcing the roots of the chicory plant, not unlike the way people force blossoms from forsythia or other flowering shrubs. Few Americans are aware of the laborious, painstaking process that is required to coax the chicory root to produce this elegant salad green. The Belgian endive is produced by hand labor without the aid of any farm machinery other than a spade or a hoe.

The Belgian farmer can't go to his seed store and order Belgian endive seeds because there is no such animal. He has to start off with seeds of the chicory plant. These seeds are sown by hand, and after about six weeks the chicory plants have roots that resemble thin white carrots. The plants are carefully dug up, the tops are removed, and the roots are planted indoors under mounds of sand and soil. These mounds provide the warmth, humidity, and total darkness required to force the roots to sprout. When the roots eventually sprout the desired white, tightly compacted leaves, they are again dug up and transplanted to beds of loam. After several weeks the cylindrical-shaped white-leafed sprouts are gathered, washed, graded, and carefully packed in ten-pound cartons. This tedious hand-labor process can't be speeded up, but once the endive is ready to ship, it is rushed to the large markets in a matter of hours.

Since as little as a single head of endive, one fifth or one fourth of a pound, goes a long way in a mixed salad, its high cost per pound is not prohibitive. When served alone, or when baked or braised and served as a hot vegetable, it is a luxury item.

When purchasing Belgian endive, look for firm, plump, crisp, unblemished, pure white heads that have yellow leaf tips. If those tips are green rather than the desired yellow, the endive isn't fresh and will taste very bitter. The short, fat heads are preferable to those that are long and thin.

Radicchio is a red-headed cousin of the Belgian endive. It looks like the heart of the red cabbage and is about the size of a small head of lettuce. There are several varieties that are used in Italy, but the one they export to the United States is called Red Verona chicory. It has a very pretty purple-red color and the leaves have a nice, smooth, string-free texture. Radicchio has almost the same bitter flavor as the Belgian endive, but it isn't quite as crisp. It is now in vogue as a salad item and can be found in high-priced specialty shops. Radicchio is very similar in flavor and texture to the Belgian endive, which sells for half the price. As it becomes more popular, increased supplies of radicchio will eventually translate to lower prices. Until then, unless red is your favorite color, the Belgian endive is a better buy.

Curly endive (curly chicory on the East Coast) is loosely headed with curly, ragged-edged leaves. The heart of the head is yellow and the outer leaves are green. It is usually used as a salad green, but in Europe it is used primarily as a cooked vegetable and in making soups. It is very crisp when fresh and has a rather bitter flavor. Curly endive usually sells at moderate prices. It is fairly perishable and should always be refrigerated.

Escarole is a flat-leafed cousin of curly chicory and has a similar bitter flavor. The leaves are thicker, flatter, and greener, but it also has a yellow heart. It is also used in salads and as a cooked vegetable. Treat it and use it as you would curly chicory. Both chicory and escarole, when limp, can be revived in ice cold water.

Watercress is an aquatic plant grown along stream banks. Its small green leaves have a sharp, pungent, peppery flavor. Watercress is used as a garnish as well as in salads. In England, no afternoon tea is complete without watercress sandwiches. However, it is very perishable. Purchase it only if it looks very green and fresh and store it in your refrigerator. Watercress is always sold in small bunches and usually at moderate prices.

Salad Fixings

❧

ANISE (SWEET FENNEL)
·
AVOCADOS
·
CELERY
·
CUCUMBERS
·
RADISHES
·
SCALLIONS

T his section includes a wide variety of botanically unrelated items that are grouped together because they are primarily used in fresh salads. Included in this group are avocados, which grow on trees; celery, which grows in muck; anise, which is related to celery; cucumbers, which are edible gourds; scallions, which are green onions; and radishes, which are members of the turnip family. Tomatoes and peppers, which are both members of the nightshade family, are listed on pages 168 and 178.

All these salad fixings are available twelve months of the year and, except for avocados, should be refrigerated immediately after purchase.

ANISE (Sweet Fennel) 🍃

Anise, which is also marketed as sweet fennel, is related to celery. It originated in southern Europe and is highly prized in Italy and France as well as in India and China, but it is only now gaining popularity in the United States. Once found only in Italian neighborhoods, it is now often available in supermarket chains that have upgraded their produce departments.

Anise is in season from early fall to early spring. It doesn't do well in hot weather and is usually out of season in the warmer months. California is the major producer, but a fair amount is grown in New Jersey.

Anise is a most unusual-looking plant. The very attractive, delicate, fernlike green foliage resembles and is often mistaken for fresh dill weed. These greens are usually discarded but may be chopped up and used sparingly as a seasoning. The white stalks of the plant aren't often used because they tend to be hollow and pithy. The choice parts of the plant are the large white bulbs that grow aboveground. They have a texture similar to celery, but are not quite as stringy and have an unusual, pleasantly sweet, licoricelike flavor. Serve anise as you would celery hearts, along with olives. They make an interesting and unusual cooked vegetable. When selecting fennel, look for fresh green foliage and crisp, firm white bulbs. If the foliage is yellow or limp and the bulbs look dry and are discolored, wait for better quality and buy celery instead.

AVOCADOS

The avocado is a tropical fruit native to Central America. Early in the sixteenth century it was introduced to Europe by the returning Spanish explorers. Today, avocados are grown in all areas of the world that have frost-free climates similar to that of their original habitat.

The United States is the world's number one commercial producer of avocados. It is a major cash crop in Southern California and southern Florida, and to a much lesser degree in Texas. California has the lion's share of the production, about 80 percent, and their avocados are available twelve months of the year. The Florida avocados have an eight-month season. They are not available during the months of March, April, May, and June. Whether your local supermarket offers California or Florida (or both) avocados depends on the time of year and your geographical location.

The avocado has a unique flavor and texture. All the other tree fruits have either a tart, tart-sweet, or sweet flavor and a juicy texture. The avocado has a bland flavor and a buttery texture. The avocado looks like a huge green olive and, like the olive, has a single hard pit. It is very firm when immature and is rich in oil when it reaches full ripeness.

There are at least two dozen varieties of avocados grown commercially in the United States. They come in assorted sizes and shapes. One California avocado is petite, weighing only a few ounces, while some Florida varieties can weigh as much as three pounds.

Depending on the variety, the immature fruit comes in every possible shade of green. Some are smooth and shiny, others are dull and have pebble-grained skins. Some varieties retain their original green color as they ripen. In others, as the fruit ripens the green changes to bronze, reddish purple, or even jet-black. Some varieties are almost round, but for the most part avocados are pear-shaped. Hence they are often called avocado pears.

Nearly all other tree fruits have to be harvested at a certain point of maturity lest they get too ripe to ship to market or even for immediate consumption. However, the avocado never reaches full maturity unless it is severed from the tree. In some California varieties the harvest can be delayed for months on end without affecting the flavor or the quality of the fruit. This ability to warehouse the fruit right on the tree is a boon to the growers because it provides for an orderly flow to market and extends the length of the season.

There are two distinct strains of avocados. The varieties grown in California are offshoots of the original Mexican and Guatemalan avocados. Those grown in Florida are derived from the West Indian avocados. Since the soil, amount of moisture, and climate of Southern California differ from that of southern Florida, the varieties that thrive on the West Coast don't do nearly as well on the East Coast, and vice versa.

While the avocado from either area is a quality product, there are significant differences in size, texture, and flavor. The Florida avocados offer advantages in size and often in price. They are usually at least twice as big as those from California and nearly always less costly. The smaller, costlier California avocados have more of the desired nutlike flavor and a richer, creamier texture than the more watery Florida fruit. A California avocado is to a Florida avocado as ice cream is to ice milk. However, if you prefer ice milk, you may also prefer the Florida avocado because of its lower calorie count.

At full ripeness, the California avocado is not quite as perishable as the fully ripened Florida fruit. A very ripe, unbruised California avocado usually cuts fine and shows no discoloration. A very ripe, unbruised Florida avocado sometimes cuts dark.

Until very recently, avocados sold at the retail produce counters were always very firm. The newest wrinkle in marketing avocados is to offer those that are ripe and ready to serve, a concept that has been promoted by the California Avocado Commission. Participating retailers, including most of the major retail food chains, report significant increases in sales. However, there is probably a comparable increase in the number of bruised, discolored, and even black avocados that have ended up in consumers' garbage cans.

I advise against the purchase of ready to serve avocados because the ripe fruit is far too fragile to survive the rigors of a self-service method of marketing. They are too often thumbed and squeezed until they are badly bruised by well-meaning but overzealous shoppers. The bruised area of an avocado will discolor and have to be cut away.

I recommend purchasing firm avocados and ripening them at home by leaving them at room temperature for a few days. If you want to speed up the ripening process, put the avocado in a brown paper bag along with a fresh tomato and put the bag in the warmest area of your home. This combination of warmth and the natural ethylene gas exuded by the avocado, and especially by the tomato, could cut the ripening time in half.

To test for ripeness, cradle the avocado in the palm of your hand. If it yields to the slightest and gentlest pressure, it is ready to serve, if it is a Florida avocado. If it is of the California variety, give it an extra day. Too many avocados are cut and served before they have reached full maturity and flavor. Once the fruit is cut, the ripening process is terminated. So make sure that it does have the slight yield before you cut it.

Avocados are not only flavorful, colorful, and nutritious but are also blessed with versatility. They can be sliced, diced, puréed, or served on the half-shell. They are flavorful enough to serve alone, but they also blend well when served with fresh fruit, salad greens, cottage cheese, cold meats, and especially seafood. A fully ripe avocado has the consistency of soft butter and makes a delicious and colorful sandwich spread. The increase in the popularity of Mexican foods has increased the usage of avocados. Their bland flavor helps take the sting out of the fiery dishes.

A cut avocado, like a sliced peach or banana, will darken and discolor when exposed to air. Sprinkling the exposed surfaces with fresh lemon or lime juice will retard this discoloration. Try to use a cut avocado as soon as possible. In the interim, cover the exposed surfaces with plastic film. If you cut the avocado in half, don't remove the pit until ready to serve.

Avocados are tropical fruits and don't like cool temperatures. Never put a firm avocado in your refrigerator. At best it won't ripen properly, at worst its flesh will turn black. A black-skinned avocado, however, is a hallmark of quality. The California Hass variety is an ugly duckling that has a dull, pebble-grained green skin when it is immature. As it ripens, the color of the skin turns to jet-black. This least attractive variety is by far the finest-flavored avocado available. When you see this Hass variety, remember that its ugliness is only skin-deep.

Never freeze a whole or cut avocado, but avocado purée may be frozen. Even though there are many recipes that call for cooking avocados, the peak flavor and texture can only be found in the raw fruit.

Before World War II, the avocado was sold only in elite fruit shoppes at costly prices. Today it can be found in most supermarkets, selling at moderate prices. Annual increases in acreage and crop yield augur ample supplies and even more modest prices.

CELERY

Celery, which is related to carrots, parsnips, parsley, and anise, originated in the Mediterranean area, where it can still be found growing wild in marshy areas. While most fruits and vegetables originated in the Far East and Middle East and worked their way westward, celery made a reverse commute and traveled eastward. While fairly popular in the West, it is a major ingredient in Chinese cooking.

The original wild celery was coarse, stringy, and sharp in flavor. The stalks were hollow and tended to be woody and were too tough to eat raw. Much like the tops found on celeriac (knob celery), the celery of yore was used only to flavor soups. Today's highly improved cultivated varieties have had the sharp flavor and tough texture bred out. Now celery is sweet, crisp, tender, and not very stringy. Although it makes a good cooked vegetable, especially when braised, it is most commonly served raw as celery sticks or celery hearts and in mixed salads. The tiny celery seeds (a million weigh less than a pound) are used in making pickles.

Celery is available twelve months of the year. Our largest and best source is California, but it is also a major cash crop in Florida, New York, Michigan, and Ohio. California ships year-round; Florida doesn't ship in the summer months, when both New York and the Midwest do. The poorest time of year for celery is late spring. That is when the California crop starts to go to seed. The tender hearts (the best part of the celery) start to solidify and will eat woody.

While there are many varieties of celery, there are two basic types and they can be identified by color. The all-green celery is called Pascal and the yellow or white varieties are called Golden. Nearly all the celery sold today is of the Pascal variety, yet fifty years ago it was the other way around.

The Pascal celery is thick-ribbed and almost string-free, and has a sweet flavor and a fairly long shelf life. Thanks to these virtues it has captured a 99 percent share of the market. The Golden, or white, celery has thinner ribs and a sharper flavor, is more stringy, and gets limp in a few days—which explains its demise.

Select crisp, clear-stalked bunches; avoid those that are limp or scarred. (Limp celery may be brought back to life if immersed in ice cold water for a few hours.) Celery must be stored under refrigeration.

CUCUMBERS

Cucumbers are related to squash and melons. They are produced in all states and are available twelve months of the year. During the winter months supplies from California, Florida, and Texas are supplemented by imports from Mexico and the islands of the Caribbean.

There are many varieties of cucumbers, but they can be broken down into three basic types: the run-of-the-mill, smooth-skinned garden cucumbers; the small, warty-skinned pickles; and the elongated, almost seedless European cucumbers.

The regular cucumbers can be found in all retail outlets and usually sell at modest prices. They are used primarily in mixed salads but are used occasionally in cold soups. They may be hollowed out and stuffed with rice and ground meat, using the same recipe as you would for baked stuffed peppers.

Garden-variety cucumbers are shipped to market in four grades, but the consumer is not privy to this information. All grades of cucumbers usually come from the same field. They are sorted and graded at the packing houses and the grades are stamped on the shipping containers. The retailer is not required to post these grades. The best cukes are graded as super select or small super. To earn this top ranking the cucumbers have to be clear-skinned, dark green in color, straight and symmetrical in shape, and fairly thin in diameter. The next best grade is called select. These are not quite as symmetrical and may be slightly off-color.

The cukes that do not meet the super select or select standards are graded as commercial and the oversized, very thick cucumbers are graded as slicers. While the slicers look impressive, they usually sell at wholesale for less than half the price per bushel of the super selects. The slicers are no bargain because they are sometimes puffy rather than firm and they always have very large, coarse, oily-flavored seeds. The smaller the cuke, the smaller the seeds, and the better the flavor.

The best-flavored cucumbers, but not the best-looking ones, are called pickles or Kirby's. The Kirby is a variety that is pale green and off-white in color. They are quite small, but the smaller the better. They aren't very symmetrical and the skin is warty. However, in this case looks are deceiving. They are crisp, crunchy, and the tiny seeds give them a fine flavor. The Kirbys are also used to make the dill pickles that you buy in jars.

The smallest variety of pickle is called the gherkin. These are also very tasty when eaten raw, but they are seldom found in retail markets. The commercial processors contract for them by the ton and pay the grower top dollar.

The European cucumbers rate an A for color, size, and shape, but not for flavor. They are practically seedless, but it is the seed of the cucumber that carries most of its flavor. Because they lack seeds, they are sometimes marketed as burpless cucumbers. While they can be grown outdoors on trellises, they are usually grown in greenhouses. They are always overpriced when compared to the price of the regular cukes.

When choosing any type of cucumber, the darker the color (except for the Kirbys), the better. A yellow color indicates old age and possibly hard seeds. The slim medium or small cucumbers are preferable to the big fat ones. Avoid any cucumbers that are puffy, soft, withered, or shriveled (soft ones sometimes have a bitter flavor). Cucumbers keep for more than a week in the refrigerator, but it is preferable to use them in a day or two after purchase.

RADISHES

Radishes come in assorted colors, shapes, and sizes. The most common ones are of the small, red, globe-shaped variety and are marketed either in bunches with their green tops attached or in cello bags (usually six ounces) without the greens. The bunched radishes are harvested and bunched by hand, then packed with crushed ice and rushed to market. They are fairly perishable and must be sold within a few days of picking or the tender green tops start to break down. If those tops are withered, and especially if they have started to turn yellow, it's a telltale sign that the radishes aren't fresh. In some areas, these tops, provided they are fresh and green, are highly prized as a salad green.

Once you bring these bunched radishes home, clip off the greens immediately. If left attached to the radishes, they will hasten their demise. This advice also applies to carrots, beets, and kohlrabies, which are also often sold with their green tops attached.

The topless radishes in the plastic bags have a much longer shelf life. The ones you purchase could be quite fresh, but chances are they have been kicking around for a few weeks. Give the topless radish a good squeeze, but please don't tear the bag. If the radish is soft or rubbery, it isn't fresh. If it is very firm and unyielding, it is fresh.

As a rule, most self-service markets prefer to sell the packaged radishes because they require little or no care and last much longer. The bagged radishes usually sell for about half the price of the bunched radishes because they are harvested, washed, topped, and bagged by machine.

The bunched radishes are offered by the small specialty fruit stores, the roadside farm stands, and the more posh supermarkets. If you have a choice, your best bet is to pay the premium price and buy the bunched radishes, provided that the tops are fresh and green. Too often the bagged radishes aren't as fresh.

There are also some radishes that are pure white in color. The most common of these is a long, slender variety called the Icicle. They usually come topless in cello packages and are crisp and zesty when they are fresh. Use the same pressure test described above to determine freshness.

A giant white radish that looks like a huge albino carrot is also available. These unusual radishes are of Oriental origin and are called daikon. Up until a few years ago they could be found only in Oriental food stores, but today they can be found in many of the larger super-markets. Daikon is marketed without tops and isn't nearly as perishable as the smaller white or red radishes.

Black radishes look like huge, ebon-colored beets. But beneath their unattractive outer appearance lies a pure white flesh that is crisp in tex-ture and has a sharp zesty flavor. They have a long shelf life and, if stored in a cool area, stay firm and fresh for months on end. Black radishes are widely used in Eastern Europe. In America they are primarily sold in markets that have a Russian or Polish clientele. The traditional Old World way to use black radishes is to slice them into wafer-thin strips, ice the slices, and serve them with rendered chicken fat, salt, and black bread.

The ugly duckling of the radish family is the horseradish. This hard as a rock, gnarled root is very sharp and pungent in flavor. It is too hard to cut with a knife and has to be grated before it can be used. White vinegar is often added to the grated horseradish to make a rather fluid paste. Sometimes a raw grated beet is added for color and to cut the sharpness of the flavor. While commercially prepared horseradish isn't quite as sharp and zesty as the homemade product, it is an acceptable time-saving and less costly substitute.

SCALLIONS

Scallions, also known as green onions, are available twelve months of the year. The best scallions and the largest crop are produced in California, although Ohio, New Jersey, and Texas have sizable crops. They are also grown for local consumption in most states.

Most Americans eat scallions raw. They use only the white root end of the stalk and discard the top portion with the green leaves. Yet in Oriental cuisine, the green leaves are highly prized.

Scallions have greatly improved in flavor over the years. They used to be quite strong, but newer varieties are quite gentle and mild.

If the scallions look crisp, if the leaves are dark green, and if the white area isn't discolored, they are fresh. Fresh scallions last for about a week when refrigerated.

Legumes
(Beans and Peas)

$\boxed{\text{B}}$ eans and peas are legumes. They are not only used as fresh vegetables but are an inexpensive source of protein when they are dried. Peanuts are also legumes. They are primarily used as nutmeats after they have been roasted. Peanuts are covered on pages 128 and 129.

There are two types of beans: pod beans and shell beans. In the ones known as pod beans, the entire pod and the inner seed (which is called a bean) are consumed. Pod beans are harvested prior to reaching maturity. In the shell beans, only the inner bean is used and the outer pod is discarded.

Peas also come in two types. In one only the inner seed (called a pea) is used and the pod is inedible; in the other the entire pea, pod as well as seed, is used.

All legumes are available twelve months of the year and should be refrigerated immediately after purchase.

GREEN AND WAX BEANS

During the winter months the North is supplied with green beans shipped from Florida, California, and Mexico. During the rest of the year they are grown locally throughout the country.

These beans are harvested while they are still very young, when the pods, either flat or round, are tender enough to eat raw and when the inner seed (also called the bean) has just started to form. They are widely used in all areas of the country as a fresh, canned, and frozen vegetable.

In most of the nation green beans are known as string beans, dating back to when most of the varieties had an inedible fibrous string that ran the length of the bean. In the past fifty years stringed varieties have disappeared from the marketplace and been replaced by newer varieties in which these inedible strings have been bred out. In the South, green beans are known as snap beans, describing what happens when it is bent in half—it snaps like a twig. If it doesn't snap but just bends without breaking, it will eat as tough as shoe leather.

The more common varieties of green beans are the round podded types. The better round varieties include Black Valentines, Harvesters, and Contenders. All the round podded varieties are sometimes called Plentifuls. The flat beans used to be called Bountifuls but are now known as pole beans. The best of these is the Kentucky Wonder. While the pole bean varieties are usually much larger than the round varieties, those of equal quality are equal in tenderness and flavor. Beans that are yellow in color rather than green are known as wax beans. There are some green beans that reach a length of eight to twelve inches. One variety originated in France and is called Haricots Verts. A very long variety that is highly prized in Oriental cuisine is called the Chinese long bean.

Although we now have better refrigeration and faster transit than we had twenty-five years ago, today's green beans are seldom as young and tender as they used to be. This decrease in quality results from the difference between machine-harvested and hand-picked beans.

Today there is a scarcity of farm labor in the United States, and what is available is too costly to compete with the machine. Until the 1940s the vegetable and fruit crops were harvested by migrant workers working for pitiful wages and under shameful working conditions. Today they are somewhat protected by federal laws and are paid a minimum wage. Although these wages are far from sufficient, it is now too costly to hand pick some crops and migrant workers have been replaced by mechanical picking machines.

These machines are mechanical marvels but are not nearly as selective and careful as hand labor. The machines can't handle very young beans of the more fragile but better flavored and more tender varieties without breaking them. As a result the growers use less tender varieties and don't pick them until they can withstand the rough handling of the machines.

Nearly all the round beans are now picked by machine. Since the flat pole beans can't be picked by machine, they very often are more tender than the round varieties. So even if you have never tried them before, buy pole beans if the round beans are not up to par.

Identifying tender green or wax beans of top quality is a snap. If they don't snap, don't buy them. Professional produce buyers determine the quality of the bean by the way it feels. A young, tender bean will have a pliable, velvety feel. Only buy beans that feel fresh and look colorful. Avoid those that look or feel coarse and dried out or are discolored.

LIMA BEANS

The lima bean has a rather flat, wide, green pod that contains flat, green, kidney-shaped beans. As a rule the pod is discarded and only the inner bean is eaten. Because some people are allergic to them, we are often advised not to eat raw limas.

There are two types of limas: The ones that have small beans are called baby limas and the ones that have very large beans are called Fordhooks. In both varieties, the greener the inner bean, the better the flavor and the texture. White lima beans are dry and starchy. When selecting any variety of limas, choose those that have full, green, velvety pods.

At today's supermarkets fresh limas are a rarity, even though they were once fairly common. They were bumped off the produce counters by the frozen limas, since they freeze better than most green vegetables. Frozen limas are almost equal in flavor and texture to fresh limas. However, they are far easier to prepare and are often, oddly enough, less costly.

CRANBERRY BEANS

Cranberry beans are our most colorful legume. They are shaped like green string beans, but the pod is cream-colored and speckled with pink markings, as is the inner bean. The pretty outer pod is never used. In the United States, primarily in the South and in Italian neighborhoods, cranberry beans are used as a fresh shell bean. In Latin America they are used primarily as a dry bean.

When selecting cranberry beans, pick out the most colorful pods. Very often they will be offered with not too attrractive, dried-out pods. Open one up—if the inner bean is colorful and fresh-looking, they are worth buying. The off-color cranberry beans should be offered at reduced prices.

FAVA (BROAD) BEANS

Fava beans are available only a few months of the year in areas that have a representation of people with Italian, Greek, and Middle Eastern heritages.

The fava bean looks somewhat like a huge, overgrown green pea. Inside, the pale green, velvety pod is tightly packed with about six to eight beans that resemble large round limas. As with limas, the pods are edible only when they are very young and immature. As a rule, the pods are discarded. Fava beans, if available, arrive in spring and are out of season by early summer. California and New Jersey produce most of our crop.

Although I have seen some of my colleagues in the produce industry nibble raw fava beans as if they were peanuts, without any adverse effects, I advise against this practice. Some people are allergic to raw fava beans and ingestion of the uncooked favas can result in mild or acute discomfort and, in rare cases, can induce a coma. The cooked fava is not toxic.

PEAS

There are several types of fresh peas. The garden peas that are sold at the greengrocers are called English peas. In English peas the inner seed, which is also called a pea, is used and the outer, inedible, pod is discarded. Young English peas are tender enough to eat raw.

Some pea varieties with flat edible pods and very flat, minute seeds are known as jade peas or snow peas. A comparatively new type of pea, as full as the English pea and with edible pods, is called the sugar snap. Both the snow peas and the sugar snaps are tender enough to eat raw.

There are also some Southern table peas, usually sold below the Mason-Dixon Line, that look more like beans than peas. These include

Crowder peas and black-eyed peas, both of which are sometimes called cow peas. The pods are inedible and the peas are not tender enough to eat raw.

ENGLISH PEAS

Prior to World War II, fresh English peas were as commonplace as green beans in retail produce markets and usually sold at very modest prices. After having all but disappeared from the marketplace, they are now making a comeback, not as a commonplace staple but as a gourmet food. They are also selling at gourmet prices.

Fresh peas lost favor during World War II because Rosie the Riveter, after putting in eight to ten hours at work, had neither the time nor the energy to shell fresh peas or lima beans or to shuck sweet corn. Fresh peas didn't regain their popularity after the war because of the introduction of the first frozen foods. While we always had canned peas, they weren't fully accepted as a "fresh" vegetable because of their lack of color—closer in hue to gray than to the desired green. I can remember restaurants in that era adding bicarbonate of soda to restore the desired green color. The bicarb may have brought back the color, but it killed what little flavor the canned peas had. Frozen peas, while not nearly as tender and sweet as the fresh ones, did have the right coloring and were accepted by most people even though they were usually more costly than the fresh product. Ironically, today fresh peas often cost twice as much per serving as frozen peas.

Fresh peas are available twelve months of the year and the peak of season is May, June, and July, when peas are grown locally in most states. During the winter months we get peas from Mexico and California. Fresh peas are usually quite costly. Remember when figuring out the cost per serving that at best you can get only two servings from a pound of peas.

When shopping for English peas the color is most important, and that color is green. Look for glossy, bright green, smooth-skinned pods. Check out the calyx—the remnant of the blossom at the stem end—and make sure it looks fresh and green. The pods should feel velvety. Reject those peas that are off-color (yellow) or dull and limp. Especially avoid those that feel very hard and have dry, rough pods. If they are in a mass display at your market, open up a pod and examine and sample the peas. If they are sweet and tender, make the purchase. If they are hard or are starting to sprout, buy frozen or canned peas instead.

Fresh English peas are one of our more flavorful vegetables. Their flavor and texture is comparable to that of fresh picked sweet corn. Like corn, as soon as they are harvested and exposed to heat, their sugar is converted into starch. This conversion isn't quite as rapid in peas as in corn, but it is a relentless process. As in sweet corn, peas that are overgrown, or exposed to heat, or that aren't fresh, will eat tough and starchy. They can be stored in a refrigerator for a few days, but the sooner they are used the better.

Fresh, sweet English peas can be eaten raw almost as a confection. Therefore, when you cook them use as little water as possible and keep in mind that you can easily overcook peas but it's almost impossible to undercook them.

SNOW PEAS (CHINESE PEA PODS)

Snow peas are smaller and much flatter than the more common English peas. Unlike the English peas, which have inedible pods, the entire snow pea is sweet and tender. Snow peas are highly prized in Oriental cuisines, where they are often left whole or sliced into thin strips and quickly stir-fried to retain their color and crispness. Snow peas used to be sold only in Oriental food stores but today are available in most larger retail markets. They are now available twelve months of the year. During the winter months they are flown in from Central America.

Snow peas often carry expensive price tags—two to three times per pound as much as English peas—but unlike English peas, where there is at least 50 percent waste, 100 percent of the snow peas end up on the plate.

When shopping for snow peas, use the same guidelines for identifying top quality as you do for English peas. Look for fresh-looking, velvety-feeling pods that are free from mold. The color of fresh snow peas is paler than that of the darker green English peas. If the snow peas are yellowish in color, or if there is any sign of decay, or if they don't feel fresh and crisp, pass them by.

SUGAR SNAP PEAS

This is a fairly new variety. They look like miniature English peas, have bright dark green pods, and are chock full of medium-sized peas. Unlike English peas, which have inedible pods, but like the snow peas, every bit of the sugar snap is usable. They are also excellent when eaten raw. Sugar snaps are available twelve months of the year and usually sell for prices that are comparable to the going rate for snow peas. While they have yet to be accepted by the Oriental trade as a substitute for snow peas, the sugar snaps are very popular with Western customers.

The Nightshade Family

EGGPLANT

·

PEPPERS (SWEET AND HOT)

·

IRISH POTATOES

·

TOMATOES

A lthough there is little similarity in appearance, eggplant, peppers, white potatoes, and tomatoes belong to the same botanical family that is known as the nightshade.

EGGPLANT

The eggplant, a subtropical plant that originated in India, thrives in warm weather and does poorly in cooler climes. Temperatures dipping below 50°F retard its growth; a whole crop can be wiped out by the lightest of frosts. The eggplant's sensitivity to cold temperatures explains why it is a staple, almost a staff of life, in the Far East, Near East, northern Africa, and southern Europe, yet is hardly used in northern Europe. As a rule, people with Anglo-Saxon, Nordic, or Germanic heritages eat eggplants only when they dine out in Italian or Greek restaurants.

In the United States the eggplant is a big seller in the larger urban cities, but it doesn't do nearly as well in America's heartland and just dies in Dixie. A lot of eggplant is grown in the South, but nearly all production is shipped north. Our annual per capita consumption of eggplant is meager—less than one per person—but the demand is on the rise.

Eggplants are available twelve months of the year. By far our biggest domestic producer is the state of Florida, which ships eggplants to market nine months of the year. When the Florida crop goes out of season, those from New Jersey take up most of the slack. California and the Carolinas also have big crops. During the winter months huge quantities of eggplants are imported from Mexico.

Eggplants come in a variety of colors, sizes, and shapes. Most of those sold in North America are of the purple-, almost black-, skinned varieties. These eggplants are oval, pear-shaped, and about the size of a small football. Some varieties are more round than oval, and there are some eggplants grown in home gardens that are thin, elongated, and almost a foot long. The miniature varieties are known as Italian eggplants even if grown in North America. They are excellent when stuffed with cheese and meat and bread crumbs and then baked whole.

The color range of the different varieties of eggplant goes from a deep purple-black to white. The pink and pink-and-white-striped ones are available in Latin American neighborhoods. There are also chocolate-colored and cream-colored varieties. However, the skin color plays no role as to the quality, texture, and flavor.

There are both male and female eggplants, identifiable by the shape of the scar in the depression at the blossom end. If it has a round dot that looks like a dimple, it's a male; if it has a dash, it's a female. At one time the male eggplant was more desirable because it had fewer seeds than the female eggplant. However, today's newer varieties, male or female, have very few seeds.

It's easy to identify fine eggplants. Choose those that are firm but not hard, unscarred, glossy, and symmetrical in shape, and have green caps. This green cap or calyx is a telltale clue to freshness. If two eggplants are of equal size, the one that is lighter in weight will have fewer seeds. Avoid eggplants that are dull in color, wrinkled, and soft, and have brown rather than green caps.

It is unfortunate that eggplant doesn't play a more important role in the American diet. It can be a flavorful meat substitute as a main-course dish, and it is usually very inexpensive. However, it is one of the few vegetables that should never be eaten raw.

☐ WHEN TO BUY: *Available year-round*
☐ WHAT TO LOOK FOR: *Firm but not hard, unscarred, glossy vegetables with green caps*
☐ HOW TO STORE: *Require refrigeration*

PEPPERS (SWEET AND HOT)

Peppers are native to the Americas and were unknown in Europe and Asia until the sixteenth century. Today peppers are grown all over the world and play an important role in most cuisines. They also are available twelve months of the year. Since Florida, California, and Texas have the longest growing seasons, they are the top three producers in the United States. However, New Jersey, North Carolina, and some of the mid-western states also have substantial crops. Peppers are easy to grow and have a high yield, so they are a favorite with home gardeners. During the winter, especially if the Florida crop is damaged by frost, we import substantial quantities from Mexico. Barring extremes in weather, peppers are in ample supply year-round and sell at moderate prices.

The fresh peppers that you purchase at the produce section of the supermarket are in no way related to the ground or whole black or white pepper on the spice rack. The tinned black pepper is the dried berry of a shrub that grows in Southeast Asia. However, the powdered paprika on the spice rack is made from sweet red peppers, and the ground red Cayenne pepper is made from hot red peppers. Tabasco sauce is made from hot red Tabasco peppers.

Peppers come in assorted sizes, shapes, and colors, but only in two flavors: sweet and hot. Depending on the variety, the size of peppers can range from as small as a thimble to as large as a small grapefruit.

Some of the large varieties are shaped like a bell and are called bell peppers. Those with blunt ends are known as bull-nosed peppers. The one with ends that taper are sometimes called sheep-nosed peppers. The small round ones are known as cherry peppers.

Depending on the maturity as well as the variety, peppers range in color from dark green, pale green, golden yellow, and pale yellow to dark red, light red, and dark purple (almost black). Some are variegated, part red and part green.

Like its cousin the tomato, when a pepper is harvested prior to reaching full maturity it is grass-green. Had that same green pepper been left on the plant for another twelve to fourteen days, it would have been red in color. The redder the pepper, the sweeter the flavor, but the softer the texture and shorter the shelf life as well. Green peppers are crisper and not as sweet. Because red peppers are more perishable, they usually sell at higher prices.

You won't be able to judge whether a pepper is sweet or hot by its skin color. The amount of heat is determined by the variety. Some are fiery, some are mildly hot, and some are sweet. As a rule, the larger peppers are mild and sweet and the tiny peppers are hot, but there are exceptions. The smaller the hot pepper, the more fiery it is.

Some peppers feature thick, fleshy walls and are very firm. Other varieties have thin walls and are not very firm.

The most common and popular variety of sweet pepper is called the California Wonder and it is grown all over the world. This is a large, thick-walled, bull-nosed bell pepper that has a deep green color. When allowed to mature on the plant, the deep green color gradually changes to a deep red. Nearly all the green peppers sold in the market are California Wonders or varieties that closely resemble it. They are excellent when used raw in salads and are an ideal size and shape for stuffing.

Another fairly common variety of sweet pepper is known under several names: Cubanelles, Italianelles, or frying peppers. These are slender, light green, or light yellow, thin-walled peppers that are very tender and are ideal for sautéeing. Occasionally they are allowed to mature fully and are marketed as red frying peppers.

There are countless varieties of hot peppers; they come in assorted shapes and sizes and can be either red or green in color. As a group they are known as chili peppers and are widely used in Mexican cooking. Some of the Mexican varieties include jalapeños, Cayennes, Pullas, and Serranos. The small Japanese-type hot peppers are called Hontakas and the larger variety is known as Sontankos. All hot peppers should be used sparingly and handled gingerly. Some are so fiery that unless you wear rubber gloves when preparing them, you will get severe burns on your hands.

In recent years during midwinter and early spring, some superb peppers have been flown in from Holland. They are grown under glass in greenhouses and are as colorful as they are flawless, looking as though they were carved out of wax. They come in three colors: a flaming red, a golden yellow, and a deep purple. I have never seen more attractive peppers. They are costly but, compared to the price of poorer quality domestic peppers in the market in the off-season, they are worth the price.

It is very easy to identify fine peppers. If they look beautiful, they are of top quality. When shopping for bell peppers, look for those that are very firm, smooth, bright, and colorful. Frying peppers won't be as firm, but they should be as flawless. Don't accept soft, withered, spotted, or cracked peppers.

☐ WHEN TO BUY: *Available year-round*
☐ WHAT TO LOOK FOR: *Firm, smooth, bright, and colorful vegetables*
☐ HOW TO STORE: *Require refrigeration*

IRISH POTATOES

The Irish potato is a tuber that originated in South America. It reached Ireland by way of Europe some two hundred years after it was first introduced to the Old World by the returning Spanish explorers. Early in the eighteenth century it was brought to the English colonies in North America by settlers arriving from Ireland, which is why we call it an Irish potato.

Irish potatoes are not true potatoes, nor are they related to sweet potatoes. The true potato is the yam. The sweet potato is botanically a member of the morning glory family. The Irish potato is a member of the nightshade family, which includes eggplants, peppers, and tomatoes.

Even though it is neither Irish nor a potato, this tuber is produced in greater quantities throughout the world than any other fresh vegetable. In North America it ranks number one in tonnage consumed.

While there is only one crop per year, potatoes are harvested in the spring, summer, and fall and are available in ample supply year-round. Those harvested in the fall are put in storage, with a minimum of loss to shrinkage or decay, in quantities sufficient to overlap the next year's crop. As a rule, potatoes sell at moderate and even modest prices.

The freshly dug potatoes that go directly from the field to market in the spring are called new potatoes. Those that are put into storage after they are harvested in the fall are called old potatoes when they arrive in market. The new potatoes have a somewhat different flavor and texture than the old ones. The new potatoes have a fresher flavor and are firmer and moister. While they don't fry or mash well and take forever to bake, they are excellent when boiled. Traditionally, boiled new potatoes are served with their jackets intact and garnished with a sprig of parsley.

These outer jackets of the new potatoes can be either white or red in color. Even though both are similar in texture and flavor, most people prefer and identify new potatoes by their red skins.

But note that not all red-skinned potatoes are new potatoes. Some red-skinned potatoes are put in storage in the fall, and when they reach the market in the winter and early spring they are mistaken for new potatoes. Some areas of the country, mainly Minnesota and the Dakotas, grow a red-skinned variety called Red Bliss. These Red Bliss are put in storage in the fall and are old red-skinned potatoes by the time they are shipped to market several months later. Often they are dyed with a red food coloring to enhance their appearance and are then mistakingly accepted as new potatoes. These old, round, red-skinned potatoes are no different than the old white-skinned potatoes in the market selling for about half the price.

For the record, the true red-skinned new potatoes that are grown in Florida, Texas, Arizona, and California don't arrive in market until February and are of superb quality. Any red-skinned potato sold during the months of November, December, and January are probably old potatoes that were grown in the North and are not the real McCoy. Even some sold in February, March, and April, when real new potatoes are available, could be old red-skinned potatoes masquerading as new. Since the real thing and the impostors look very much alike, the consumer will have to depend solely on the integrity of the retail market to get a fair shake.

There are countless varieties of potatoes that come in assorted sizes and shapes. They also come in a yellow flesh color as well as the traditional white. Almost all the potatoes sold in North America are white-fleshed. In recent years some of the very exclusive fruit stores have been trying to introduce a yellow-fleshed variety, called Finnish potatoes, that carry very high price tags. It is still too early to assess the success of this venture.

Potatoes come in two different shapes: long and round. Although all shapes may be used for all purposes, as a rule the long ones are used for baking and are called bakers. The round ones are known as all-purpose potatoes but are not usually used for baking.

The bakers are primarily grown west of the Mississippi. Some have russet-colored skins, others have clear white skins. The clear, white-skinned ones are grown in California and are known as White Rose or Shafter Whites. The Shafter Whites are shipped to market in refrigerated rail cars and are some of our finest potatoes. Even though they are new potatoes, the larger ones bake well. The smaller ones are wonderful when boiled. The darker-skinned bakers are called Russets. Although they can be grown in many areas, by far the largest, possibly the best, producer of this type is the state of Idaho. Most consumers refer to the Russets as Idahos regardless of where they are grown.

The round, white, all-purpose potatoes are for the most part grown east of the Mississippi. Maine is the number one state as far as tonnage. The Canadian Maritime provinces are also major producers. These round white potatoes are usually less costly than the Russets.

Most potatoes are now marketed in five- or ten-pound prepacked see-through bags. When purchasing these bags, check out the potatoes to see that they are firm, not rubbery, and free from cuts. Avoid potatoes that are greenish in hue—they have been exposed to sunlight and will have a short shelf life. If you are shopping for all-purpose eastern potatoes, the prepacked bags are the most economical way to buy them. The wide range in size won't pose a problem if you use these potatoes primarily for boiling or frying.

If for the most part you use them for baking, the prepacked western potatoes may not fill the bill. Often, the ideal baking-sized ones are skimmed off before the potatoes are bagged up. Most of the packages of western potatoes contain an excess of potatoes that are undersized for baking. You'll do better to pay a premium price and purchase individual, uniformly sized bakers. Since they will be almost identical in size, they will bake evenly.

Do not wrap potatoes in foil prior to popping them in the oven. A foil-wrapped potato retains rather than exudes moisture and when baked will be wet and even soggy rather than dry and fluffy. A potato wrapped in foil will be steamed rather than baked.

☐ WHEN TO BUY: *Available year-round*
☐ WHAT TO LOOK FOR: *Firm, unbruised vegetables*
☐ HOW TO STORE: *Refrigerate new potatoes; do not refrigerate old potatoes unless temperature exceeds 70°F.*

TOMATOES

Tomatoes, which are native to South America, are available twelve months of the year. In the United States, California and Florida are the major producers, although during the summer months tomatoes are grown locally in most of the other states. During the winter months tomatoes are imported from Mexico in huge quantities, and in far lesser amounts from Europe and Israel, to supplement our own production.

Originally tomatoes were yellow in color, but these are now an oddity—more than 99 percent of the ones produced in the United States are red. Along with the traditional-sized and round-shaped garden varieties of tomatoes, there are some types that are plum-shaped. There are also miniature tomatoes, not much bigger than marbles, called cherry tomatoes, which have been showing an annual gain in popularity.

Tomatoes were first introduced in Europe as curios by the explorers returning from the New World. In many areas of Europe it was believed that the tomato was poisonous because it was of the same botanical family as belladonna, a known poison.

This unwarranted fear was brought to North America by the early settlers, but it wasn't until the late nineteenth century that we accepted the tomato as a wholesome, flavorful, nontoxic, fresh food. Today the tomato ranks third in the annual tonnage of fresh vegetables used in the United States. (Potatoes rank first and lettuce, second.)

There has always been some dispute as to whether the tomato is a fruit or a vegetable. Since it is a berry it is botanically classified as a fruit. However, in the marketplace it is classified as a vegetable by the consumer, and the customer is always right. So we'll call it a vegetable.

Depending on your geographical location and the time of the year, you will be offered two kinds of tomatoes: vine-ripened and force-ripened. They'll look alike but won't taste alike. Tomatoes are at their flavor best when they are harvested after they reach full color. They will lack flavor if harvested while still green and then force-ripened. Let's call the full-flavored ones super tomatoes and those that lack flavor supermarket tomatoes.

The super tomatoes with full flavor are those that aren't picked until they reach full color. But even those that are picked while only starting to turn pink will ripen up to near-full flavor. Tomatoes of this quality are like the ones you would grow in your own backyard and are usually available when grown locally during the summer months. They are very delicate and fragile and must be picked, packed, transported, and marketed with utmost care to prevent bruising or crushing.

Some vine-ripes are available during the off-season in the winter months, but most of the domestically grown supply is down to a trickle because nearly all the Florida and California growers prefer to sell the consumer the green rather than the ripe tomato. In the winter some vine-ripes arrive via truck from Mexico and only a few come in via air from Europe and Israel. Prior to the energy crunch we had a fair amount of hothouse tomatoes, but this supply has been decimated due to the high cost of heating the greenhouses, most of which were located in Ohio.

The supermarket tomato is a vegetable of the same color, but it has a different flavor. These are picked while still grass-green. The growers claim that they are "mature" greens, but green is green. Green tomatoes are harvested, washed, graded, and packed by machine. Either at the source, in transit, or upon arrival at the wholesale terminals and chain-store warehouses, they are exposed to warmth and usually to ethylene gas. This combination of heat and gas force-ripens the tomatoes. They change in color from green to pink to a rosy red. At this point they look very much like the flavorful super vine-ripes, but they lack flavor and are often too hard to be juicy. Note that the ethylene gas used to ripen the tomatoes is the same product used to ripen bananas. It is harmless and is a duplicate of the natural gas exuded by many fresh fruits, including tomatoes.

To the growers, force-ripened tomatoes mean a lot more cash for their crop because the labor costs are only a fraction of those needed to produce vine-ripened tomatoes. There is also a far greater yield per acre because there is less chance of damage in the field from rain, wind, or hail, and there is little or no loss due to damage in transit. Supermarkets love the force-ripened tomatoes because they are almost indestructible. They require no extra help or care. Even if they are handled like coal by careless employees or overzealous self-service shoppers, the supermarket tomatoes endure. Similar treatment to vine-ripes would result in a transformation to ketchup or tomato juice. But during the winter months especially, consumers have to settle for less than flavorful tomatoes.

Unfortunately, even at the peak of season during July and August, when so many areas have locally grown tomatoes that are readily available, most of the supermarkets and many smaller stores continue to handle the same flavorless tomatoes they offered all winter. But at least during the summer the consumer can shop for homegrown vine-ripened tomatoes at road stands or in some of the more specialized fruit stores.

Whenever the vine-ripe tomatoes are not available, try switching to cherry tomatoes or plum tomatoes. They usually have more flavor than the force-ripened tomatoes.

Tomatoes don't like and do not ripen properly or at all (depending on the maturity of the tomato) if stored under refrigeration. Unfortunately, while there is a vast difference between the flavor of vine-ripened and force-ripened tomatoes, they look very much alike.

☐ WHEN TO BUY: *Available year-round*
☐ WHAT TO LOOK FOR: *Vine-ripened, full-color fruit*
☐ HOW TO STORE: *Ripen at room temperature; refrigerate only after fully ripened and only if temperatures exceed 75°F.*

The Onion Family

🍆

GARLIC

·

LEEKS

·

DRY ONIONS

·

SHALLOTS

O nions are our most pungent vegetable. Some are quite mild while others are strong in scent and flavor. They are primarily used in cooking to flavor other foods but are also used raw in salads. Those that are known as dry onions are bulbs that are harvested after the greens have started to decay; these greens are discarded. Green onions are types that are harvested while the greens are fresh, and the greens as well as the bulbs are used. Green onions include leeks, scallions, and chives, but only the leeks are covered in this section. Since scallions are primarily used raw in salads, they are covered on page 159. Since chives are primarily used as an herb, they may be found on page 235. All members of the onion family are available year-round.

Dry onions, garlic, and shallots are very hardy and need not be refrigerated unless temperatures exceed 70°F; unlike most vegetables, these onions have a longer shelf life if they are not refrigerated. Leeks are perishable and require refrigeration.

GARLIC

Garlic is grown all over the world. It is the most potent and pungent member of the onion family. The plant grows to a height of about twelve inches and produces delicate white flowers. However, there is nothing delicate about the bulb the plant produces. Each bulb, called a head or a knob, contains eight to twelve sections, called cloves. These cloves are covered and held closely together by a parchmentlike covering. The garlic plant doesn't produce seeds but is propagated by planting the cloves.

There are three basic types of garlic: Creole, Italian (Mexican), and Tahitian.

Garlic is believed to have originated in Asia Minor and now plays an important role as a seasoning agent in almost every ethnic cuisine, though it is more widely and lavishly used in warmer climates. This degree of usage dates back to the era prior to modern refrigeration, when garlic was used to preserve meats and to mask the flavor and odor of meat that wasn't fresh. In Sicily and Spain it is used with a heavy hand in almost all dishes except dessert. In France and northern Italy it is used more subtly. In Britain and the Nordic countries garlic plays a minor role.

Since ancient times, garlic has been credited with possessing mystical and medicinal powers. It was believed that garlic drove away evil spirits (even Dracula would have none of it). It was prescribed for everything from curing athlete's foot to restoring hair on bald pates, including all ailments between those two extremities. During the dark ages people believed that wearing a garland of garlic would ward off the plague. In America as recently as 1917 and 1918 during the influenza epidemic, people wore garlic garlands when they went out in public. Despite these age-old myths and old wives' tales, there is some truth to back up the use of garlic in folk medicine: The high supply of organic sulfur compounds contained in garlic is recognized by modern medicine to have antibacterial properties.

Most of our garlic is supplied from California, and since these supplies are supplemented by imports from the Southern Hemisphere, garlic is in ample supply year-round. Nearly all the California garlic is of the Creole variety, which features fair-sized heads, white skins, fairly large cloves, and a fairly mild flavor. The Italian (Mexican) garlic has a purplish skin and is smaller in size than the California (Creole) variety. It has smaller cloves but a sharper flavor. The Tahitian variety has an insignificant share of the total market. This variety produces extra-large heads and is often called elephant garlic. These white heads are at least twice as large as those found in the other two types. Tahitian garlic has a milder flavor and is usually sold by specialty shops or mail-order houses at three or four times the price of the smaller garlic. The extra size is not nearly worth the extra cost.

At the wholesale level, the larger-sized heads of garlic command higher prices. However, since garlic can be something of a chore to prepare, and since fewer large cloves than small ones are needed to render the same amount of garlic, larger cloves (except for Tahitian garlic) may be worth the premium price.

Shop for garlic as you would for dry onions. Ignore the white or purplish color of the parchmentlike skin; both types are of equal quality. Select firm, dry, sprout-free heads. Garlic shows age by getting soft or wet and by shooting green sprouts. It keeps for at least a month when stored in a cool, dry, well-ventilated area, but it won't keep as well in the refrigerator. Unlike the onion, which has to be used quite soon after it has been cut, you can remove as little as a single clove of garlic without decreasing the lasting power of the remaining cloves.

LEEKS

The leek is the mildest member of the onion family. Unlike the sharper onion, there isn't a tear or a sniffle in a carload of leeks. They look like huge scallions with long white stalks and flat green leaves. In North America they are primarily used as a soup green, but in Europe they are also used as a cooked vegetable. They may be, but seldom are, eaten raw.

While leeks are not a big seller in this country, they are widely used in France and Great Britain. In France, while the wealthy are dining on fresh asparagus, the less affluent use leeks as a substitute in recipes that call for asparagus. Scotland's famed and hearty cock-a-leekie soup is made from chicken, barley, and leeks. In Wales the leek is not only used as a vegetable but as a national emblem.

When purchasing leeks select those that are fresh-looking, flexible, and free from grit. If the stalk isn't pliable, it will probably be too woody. Even if the leek looks free from sand, play it safe and give it an intense washing in cold water.

DRY ONIONS

There are dozens of different varieties of bulb onions that come in a wide range of sizes and shapes and in three flesh colors: yellow, white, and red. All onions are similar in flavor, but some are sharp and others are sweet.

Yellow onions are grown in many areas of the United States, with

Texas, California, Idaho, New York, and Michigan as the major producers. In a calendar year the first onions to arrive in market are called new onions and are usually of the Grano variety. They are harvested in April and are grown in Texas, California, and Arizona. The Grano is flat-shaped and has an almost colorless, paper-thin skin. It is very mild in flavor and is one of our sweetest onions. Unlike the more hardy varieties that arrive later in the year, the Grano is fairly perishable and purchases should be limited to a few days' supply. The Granos and the other early varieties from Texas and California are in season from April until August.

The onions that arrive in market in August from the more northern states (New York, the Pacific Northwestern states, and the Midwest) are called globe onions. They are round rather than flat in shape, have copper-colored parchmentlike skins, are are known as old onions. They are sharper than the mild new onions.

When these globe onions are cured in proper storage in cool, well-ventilated areas, they have a shelf life of six to eight months. Carryovers of these stored onions from the prior year's crop are available in moderate supply when the first crop of new onions arrives in April. Since both crops overlap—barring bad weather—onions are available twelve months of the year.

Just as there are two crops of yellow onions, there are two crops of white onions. The first arrival of the white onions parallels that of the new yellow onions. Most of the white onions are of varieties that produce a very small bulb. They are primarily boiled and used in recipes that call for creamed onions and are therefore called boilers. There are also some very large white onions that are called white Spanish onions, and they are used raw. Bermuda onions are the same as Spanish onions, though they are not imported from Bermuda or Spain. Most are produced in the United States and some are brought in from Chile.

The red-skinned, red-fleshed onions are known as red onions or Creole onions. These are used primarily in salads. The early red onions are grown in Texas and California. The later varieties are grown in New York and Michigan and are also imported from Italy.

The mildness or the sharpness of an onion can't be determined by its color or outward appearance, but rather by the variety and the area in which it was produced. The seed used to grow large Spanish-type onions in Idaho produces a sharper onion when produced in New York. This is also true of the imported red Italian onions, which are sweeter than those grown in Michigan or New York. However, the red onions grown in California and Texas are quite mild in flavor and far less costly than those imported from Italy.

In recent years there has been a PR battle waged in newspapers and magazines as to which area grows the sweetest, mildest onions. The main participants in this contest are the Vidalia onions grown in Georgia, the Maui onions grown in Hawaii, and the Walla Walla onions grown in Washington State. All three are fine products of comparable quality and flavor, and they usually sell at premium prices. But the less-heralded onions grown in Texas, California, and some of the northwestern states are almost if not just as sweet and mild as those and a lot less costly.

Look for very firm, dry, well-shaped onions that are almost free from odor and completely free from sprouting. Softness at the neck (the top) of the onion is a dead giveaway as to impending or actual decay. Store onions in a cool, dry area.

SHALLOTS

At first glance the shallot looks like a very small, old yellow onion but it is actually one of the most elegant members of the onion family. The dull, copper-colored, parchment-skinned exterior hides a very distinctive flavor that is somewhere between that of garlic and onion. The shallot is highly prized by French chefs; the French word for shallot, *échalote,* is incorporated in many *haute cuisine* recipes. Bernaise and bercy sauces are not authentic when onion or garlic is substituted for shallots.

Shallots are available twelve months of the year, but as a rule those that are available in July and August aren't as good as the ones that arrive in the cooler months. Most of our shallots are produced in New Jersey and Long Island, but a fair supply is imported from France. They are quite hardy and will keep for a month or two if stored in a cool, dry, well-ventilated area.

Shop for shallots as you would for garlic or dry onions. Select those that are very firm, dry, free from sprouting, and well covered with the parchmentlike skin. And be prepared to be greeted by lofty price tags when purchasing them; the limited demand usually exceeds the limited supply.

Root Vegetables

❦

BEETS

·

CARROTS

·

CELERIAC (KNOB CELERY)

·

PARSNIPS

·

RUTABAGAS (YELLOW TURNIPS)

·

SALSIFY (OYSTER PLANT)

·

WHITE TURNIPS

N early all root vegetables are available year-round. As a rule they sell at fairly moderate prices because they are quite hardy and there is very little loss to shrinkage or spoilage either en route to retail markets or on retailers' display counters. Even though they are hardy, they have a longer shelf life if stored in the refrigerator.

The greens of most of the root vegetables covered in this chapter are inedible. The two exceptions are beet greens and turnip greens, which, if fresh, are an excellent substitute for cooked spinach. Whenever root vegetables are sold in bunches with attached greens, cut off the greens from the roots whether they be edible or inedible. If left attached the greens draw moisture from the root and drain some of the vitality and flavor.

These products are known as root vegetables because the portion of the plant used is produced beneath the earth's surface. Onions, potatoes, and yams, which are also grown underground, are discussed in other chapters.

Root vegetables are available twelve months of the year and should be refrigerated immediately after purchase.

BEETS

Fresh beets are available year-round. They are usually offered with the greens attached, banded in bunches that contain four or five roots. Although most people discard the greens, if they are young and fresh they can be cooked as a leaf vegetable. Beet greens, also known as beet tops, are similar in texture to fresh spinach. Some consumers prefer them to spinach because the beet tops won't collapse and get gooey if overcooked.

Some beets are shipped to market with the greens removed, packed in fifty-pound bags or bushels. These are primarily used by the restaurant trade. Very often these clip-top beets come out of storage and are not nearly as fresh or tender as the beets sold in bunches with the greens attached.

Although fresh beet root is one of our most colorful vegetables and has a sweet flavor and a smooth, tender texture, it is not one of our better-selling fresh vegetables.

However, back in Grandma's day beets were very popular. One of the reasons why beets have slipped in market share is probably the high price tags. In recent years the average price at retail for a bunch of beets, depending on the time of year, has ranged from fifty-nine to eighty-nine cents. Canned beets, which are quite good, sell for about half the price, although there can be no doubt that fresh beets are more flavorful than the ones in tins. The other reason for declining sales popularity is the length of time it takes to cook fresh beets. In this era, with so many women in the job market, a slow-cooking item is not acceptable to our fast-food, microwave generation.

Select fresh beets that have firm, smooth-skinned, round, dark-red globes that are free from scales or cuts. Smaller- and medium-sized beets are usually more tender than larger ones. The beets that have nice, fresh, crisp greens are preferable to those with wilted tops. However, if the roots are firm and colorful, the beets will still be of acceptable quality, even if the tops are limp and have begun to show signs of decay.

During the winter months you may be offered beet roots that are packed in bags minus tops. These clip-top beets usually come out of storage. (They have a very long shelf life if stored properly.) These topless stored beets are all right but won't be quite as tender as the fresh beets.

Beets are at their flavor and color best when they are cooked whole. To retain top color and flavor, never cut, slice, or even peel the beets prior to boiling them. When you remove the tops, always leave an inch or two attached to the root. Any exposed cut surface of the beet will bleed out into the boiling water. After the beets are cooked, the skin peels off very easily.

CARROTS

Carrots are available twelve months of the year. They are almost always in ample supply because they are not very perishable and are usually the least costly fresh vegetable at the produce counter. Despite their moderate price they are most nutritious, flavorful, and colorful, and are equally good when served cooked or raw.

Carrots are root vegetables that have green, parsleylike tops. They are in the same botanical family as parsley, but the carrot greens are inedible. Back in Grandma's day, fresh carrots in season were marketed only with the green tops attached. The clip-top carrots were harvested in the fall, stored in root cellars, and offered for sale during the winter months.

Today nearly all carrots are sold year-round without tops. They are packaged in sixteen-ounce plastic bags. But a few with greens attached are shipped in bunches containing five to eight carrots in each. The larger carrots are shipped in fifty-pound bags and are used by the purveyors who supply restaurants and institutions. In recent years miniature carrots packed in twelve-ounce bags have made their debut and are fairly popular. These are especially good when cooked whole.

When shopping for the sixteen-ounce packages of carrots, peek through the plastic and select those that contain carrots that are small or medium in size, colorful, well shaped (tapered and not blunt-ended), smooth, and very firm. Avoid those that are limp or wilted. Check out the tips of the carrots (that's the first area of decay) and check out the tops (if you see yellow sprouting, that carrot is old enough to vote).

If you purchase bunched fresh carrots with the greens still attached, clip off the tops as soon as you get home if you plan to keep them a day or two. Carrots with greens attached lose freshness faster than those without the tops. These bunched carrots sell at premium prices—usually at twice the price of the bagged carrots. While they are fresher and better flavored, the slight difference in quality does not warrant paying double the price.

Carrots aren't very perishable, but since they are so readily available and since the prices seldom fluctuate, why buy them several days before you plan to use them? Limp, rubbery carrots may be rejuvenated by putting them in cold water for a few hours.

The source of the carrot will determine the sweetness. The farther west it originated, the better the flavor. Even though most areas use the same seed, by far the sweetest carrots are produced in California and Arizona. Those from Texas, Michigan, and Florida are the next best. Carrots imported from Canada and especially those grown in the northeastern states are usually less sweet and are apt to be more woody.

CELERIAC (KNOB CELERY)

Celeriac is not one of our more popular vegetables, perhaps because it resembles an unwashed horseradish root. Yet beneath its unattractive, unglamorous shell lies a flavorful, crisp, cream-colored, smooth-textured flesh that tastes like celery. It can be used raw in fresh salads or as a cooked vegetable and it is an important ingredient in soups. While it is not much of a seller in the United States, it is very popular in northern Europe. Here at home, fresh celeriac is available every month but June and July.

When harvested, celeriac looks like coarse green celery attached to a rough-looking bulbous root. In late summer and early fall it comes to market usually tied in bunches that have three knobs and with the celerylike greens attached. The greens are too coarse to use raw as table celery, but if they are fresh and have not begun to yellow, they make a fine soup green. After the first frost the celeriac is shipped to market minus the greens and sold as a root vegetable by the pound.

Select firm medium-sized knobs; small ones have too much waste when peeled and large knobs are apt to be hollow or woody. Pressure on any darker areas of the skin will expose decay that otherwise might be hidden by the rough exterior.

If stored in a cool, moist area, such as underground in root cellars or in the refrigerator, the shelf life of celeriac can be measured in weeks and even months. When served raw, the exposed flesh tends to discolor. This discoloration can be retarded by adding a few drops of vinegar or lemon juice. When cooking celeriac, they will be easier to peel after they have been boiled.

Forty or fifty years ago, when we didn't have a wide array of fresh vegetables during the winter and early spring months, celeriac was available in most retail produce stores at very modest prices. Then it all but disappeared from the produce counters until about ten years ago, when it started to be featured in cookbooks and magazines. It is now making a slow comeback, not as a staple but as a gourmet food, selling of course at gourmet prices.

PARSNIPS

Parsnips, which look like albino carrots, are available twelve months of the year. When cooked they are as tender as and possibly sweeter than their cousin the carrot. As with carrots, only the root of the parsnip is edible, not the greens.

Parsnips are at their flavor best after they have been exposed to cold weather, and are at their poorest in midsummer. When stored in a cold, humid area or a refrigerator, they have a long shelf life.

Of the limited amount sold in North America (they don't do very well in Europe either), most parsnips wind up in stews or soups. However, they are especially flavorful when cooked along with carrots and you can prepare them as you would carrots. Even though parsnips are as easy to grow as carrots, lack of consumer demand results in limited distribution and comparatively high prices.

Judge the quality of parsnips as you would carrots. Select those that are firm, crisp, and free from cracks. Medium-sized parsnips are preferable to those that are very small or very large. Avoid those that are discolored (brown instead of cream-colored) and those that are withered and limp.

RUTABAGAS (YELLOW TURNIPS)

Rutabagas, also known as yellow turnips, Swedish turnips, or Swedes, are available year-round. It is a humble vegetable with a heart of gold, but it is surprisingly unpopular considering its interesting, earthy flavor and smooth texture. Rutabagas are probably the least expensive fresh vegetables on the produce stand and are waste free, but they just don't sell well.

Each year, when nearly all the other fresh vegetables show healthy increases in tonnage, the rutabaga barely holds its own. This worthwhile vegetable deserves better treatment.

Canada grows the world's finest rutabagas (they are at their best in cool climes). Canadian rutabagas are in season from October through July. When they are not available, we get some from California that are not as good.

Canadian rutabagas have a longer shelf life than nearly any other vegetable. To extend their durability, they are usually coated with a thin layer of paraffin that is easily removed.

After they have been peeled, the rutabagas can be cut up, boiled, and mashed just like mashed potatoes. If you find the flavor to be a bit sharp, you can tone it down by blending the mashed rutabagas with mashed white potatoes—a most compatible combination. Or, to tempt those with a sweet tooth, sprinkle the mashed rutabagas with a light dusting of brown sugar. The result will taste somewhat like baked acorn squash, but it won't be anywhere near as stringy.

When buying a rutabaga, select one that is heavy in relation to size, roundish rather than coming to a point, and as hard as a rock. Avoid any that feel the least bit soft or are light in weight.

SALSIFY (OYSTER PLANT)

Salsify is a rather unattractive root vegetable that resembles a very long, very thin parsnip. By the fullest stretch of the imagination, the flavor of salsify is supposed to be similar to that of oysters.

If we were to tally the least-used vegetable, salsify would come close to the head of the list. Before World War II, salsify was a fairly common sight in many produce markets. It was usually available early in fall and throughout the winter. Even though it was rather modest in cost, it didn't sell too well because of the length of time it took to prepare it properly. Possibly the biggest damper on its sale was the complaint that if rubber gloves weren't used in its preparation, the hands of the cook would be badly stained. Until about 1975, salsify was all but extinct. Although not available in produce stores, some was grown in home gardens. Today, salsify is making a slight comeback. Occasionally it is found in some gourmet fruit shops and in the more posh supermarkets. I've even seen some imported from Belgium and Holland. It now carries a lofty price tag.

WHITE TURNIPS

White turnips are available year-round. They are also widely grown, especially in the South, for the green tops. The root is as colorful as any fresh fruit. It is global in shape, like a beet, smooth-skinned, and pure white in color, with a royal purple crown. If they weren't so plentiful, they might well sell as an exotic vegetable.

White turnips may be eaten raw or as a cooked vegetable and are also used to enhance the flavor of soups. In the late spring and early summer months they are shipped to market packed in bunches with their green tops attached—not unlike beets. During the fall and until the new crop arrives in late spring they are marketed without the greens. In the fall these white turnips have a shelf life of months on end if they are stored in a cold area or in a root cellar.

Choose attractive-looking, colorful, firm, fresh, white turnips that are heavy in relation to size. The smaller ones will have a better flavor and texture than the larger ones. Avoid any that are misshapen, discolored, or soft. A sprouting at the crown end (top) of the topless turnip is a sign of age or improper storage.

Mushrooms
(Fungi)

CULTIVATED MUSHROOMS
·
WILD MUSHROOMS

F ungi include wild mushrooms, cultivated mushrooms, morels, and truffles as well as molds, mildews, and toadstools. Some forms of fungi are edible, others are highly toxic.

Our most common form of edible mushroom is the familiar round, white, cultivated mushroom. This originally wild variety was domesticated hundreds of years ago. (There are countless varieties of wild mushrooms. In recent years some of these have also been domesticated on a small scale.)

Wild mushrooms just seem to pop up from nowhere. They spawn unattended in fields and forests, even on lawns and golf courses. There are dozens of varieties of wild mushrooms. Unfortunately, those that are poisonous look just like the ones that are edible. Unless you are an expert (and this is a rare expertise) at telling them apart, give wild mushrooms a wide berth. Buy them in a food store. Picking wild mushrooms is no game for amateurs. Each year we lose a few "professionals" who guessed incorrectly.

Just as we farm shrimp and salmon in the sea and catfish in ponds, we have also started to farm "wild" mushrooms by duplicating conditions where they are apt to spawn. Some varieties refuse to be domesticated, especially the morels and the truffles, though it is not for lack of trying.

Some of the wild mushrooms and the domesticated "wild" mushrooms sold in retail markets include chanterelles, Enokis, oyster mushrooms, Creminis (Italian Browns), Ceps, and Portobellos. Depending on the season and the variety, they retail from eight to twenty dollars per pound. But this price is nothing when compared with the retail price of morels and truffles, which sell for thirty-five to several hundred dollars per pound. As a rule, these high-priced fungi are bought by the *haute cuisine* restaurants. However, more and more consumers are starting to purchase the more exotic fungi. At New York's Hunts Point wholesale produce terminal, several varieties of wild mushrooms are available on any given day.

Cultivated mushrooms are available twelve months a year. Supplies of wild mushrooms are erratic and subject to local demand. Both should be refrigerated immediately after purchase.

CULTIVATED MUSHROOMS

Those white mushrooms sold in your market are related to a wild variety that grows at random in the fields. These once wild mushrooms have been tamed and improved and are scientifically monitored and nurtured to achieve maximum quality and yield.

In Taiwan and China cultivated mushrooms are produced out of doors; in Europe and the United States they are grown in cellars, caves, and old mines, but mostly in specially constructed mushroom houses. They do best in cool weather, and have problems when the mercury exceeds 60°F. Prior to World War II and prior to mechanical air conditioning, cultivated mushrooms were in the market only during the nine coolest months. In the heat of the summer, mushrooms were out of season. Today mushroom houses are air-conditioned during the summer and it is now a twelve-month crop. However, those produced in naturally cool weather are usually of better quality than those grown under artificial conditions. Pennsylvania produces more than 50 percent of the cultivated mushrooms in the United States; California and Michigan are also top producers.

Until about twenty years ago, nearly all mushrooms were shipped to retail markets in oval wooden baskets that held three pounds net weight and were weighed out by the retailer at the time of purchase. Today, in addition to the three-pound baskets, a few are packed in bulk in ten-pound cartons, and large quantities are shipped to market in prepackaged twelve- and sixteen-ounce paperboard trays that are wrapped with plastic film. However, since the bulk mushrooms command a higher price than the prepacked, the finest mushrooms are packed in the familiar three-pound oval baskets.

Although most cultivated mushrooms are snow-white, some varieties are cream-colored and others are as brown as chocolate. Consumers in the East and Midwest prefer the white mushrooms; ivory-colored ones are usually sold to commercial canners; and the chocolate-colored mushrooms, a variety called Hawaiian Browns, are very popular in California. Attempts to introduce the brown mushrooms in the East have met with little success. If mushrooms are fresh and of equal quality, one color is just as good as another, though brown mushrooms have a more prominent, earthy flavor.

Mushrooms are quite perishable and are graded at the shipping point based on size, not quality. The larger- and medium-sized mushrooms are called Specials. The smaller ones, below the size of a quarter, are called Fancys. The very small ones are called Buttons. Buttons are seldom shipped to the retail stores because the canners pay a premium price for them. The size has no bearing on the quality, flavor, or texture of mushrooms. If they meet the following criteria, they are of top quality.

The perfect mushroom is pure white, pure ivory, or pure dark brown. It feels firm and dry to the touch. It is as tight as a drum as well as very firm and has a very short stem. The stem is edible but isn't nearly as tender and flavorful as the cap of the mushroom. Therefore, the shorter the stem, the better the mushroom. Reject mushrooms that are discolored, or limp and have opened up like umbrellas. Avoid purchasing those that feel sticky or tacky and especially those that feel slimy.

Mushrooms that resemble opened parasols, or that are discolored, are not fresh. At the wholesale level they sell at deeply discounted prices. If they are open and discolored but feel dry to the touch, they are still worth using, but they should be offered at sharply reduced prices. However, if they feel sticky or tacky, they may be used only if they have been washed and peeled. If they are slimy to the touch, they should be discarded.

The difference between a superb mushroom and one that should be thrown in the garbage pail is only a matter of time. A mushroom is at its best only if no more than a day or two has elapsed since it was harvested. About three or four days after harvest they start to open and discolor. In six or seven days after harvest the caps open fully and they discolor badly and start to get sticky. After another couple of days they are candidates for the garbage can. Therefore, don't purchase mushrooms unless you plan to use them in the next day or two.

Fresh mushrooms don't need to be peeled. Peeling a mushroom is not only a waste of time, it is also a waste of mushroom, since most of the flavor and some of the nutritional value is closest to the skin of the mushroom cap. If a mushroom isn't fresh and is sticky to the touch, it should be peeled. If it feels slimy, don't bother to peel it, just chuck it out.

Although the finest mushrooms are shipped to market in the three-pound baskets or ten-pound cartons, most of the self-service super-markets prefer not to handle these bulk mushrooms because they are too fragile to withstand the rigors of a self-service method of marketing.

The moment a white mushroom is handled, it starts to discolor. The self-service stores prefer to sell mushrooms that have been pre-packaged by the growers. These prepacks can absorb a lot of abuse by well-meaning but overzealous customers and do not have to be weighed at the time of purchase.

There are two types of prepackaged mushrooms. Though they sell at comparable prices, there is quite a difference in quality. Unfortunately, those that look the best are of the poorest quality.

The prepacks can be natural mushrooms, similar to the ones sold in bulk. They are packaged and wrapped immediately after harvest. However, the best of the natural mushrooms are skimmed off the top and sold in bulk. It is usually the ones that are not quite up to top quality that are prepackaged.

Most of the prepacks (fortunately their share of the market is on the decrease) are chemically washed and bleached prior to packaging. The chemical used in this process is called sodium bisulfite and cosmetically it does a fine job. Some dingy-looking mushrooms that most consumers wouldn't consider buying are transformed into an attractive snow-white product.

Natural mushrooms, provided they are fairly fresh, are a far better buy than any chemically treated mushrooms for two reasons. The first is that the chemical masks impurities and hides imperfections. A natural mushroom exposed to air and warmth won't get sticky and tacky for four to five days, even after the package has been opened. A chemically washed mushroom will get sticky and tacky almost overnight, either in or out of the package. The second reason is of much more serious consequence. The sodium bisulfite, although it is used with the blessing of the Food and Drug Administration, is highly suspect and is now being reexamined by that agency. This bisulfite was also used by restaurants to prolong the longevity and prevent discoloration of preprepared salads. Due to complaints by consumers of shortness of breath and nausea, and one report of a fatality, the FDA has put the use of sodium bisulfite on salads on hold until further testing. However, no such directive has yet been issued for its use on mushrooms. The bisulfites especially cause trouble to people with respiratory problems. With hope the very slow-moving FDA will ban the use of sodium bisulfites on all food products. Until they do, it is up to the informed consumer to read the label on the prepackaged mushrooms carefully and to buy only those that have not been treated with sodium bisulfite.

It may require twenty-twenty vision plus a jeweler's loupe to read the fine print. However, it is required by law that the consumer be alerted about any chemical additives. Unless you select the prepacks that are labeled NATURAL, even though they are not as snow-white as the chemically treated ones, you will get a snow job. This problem won't exist if you buy mushrooms in bulk; these as a rule are not treated with chemicals.

The foolproof way to recognize chemically washed mushrooms is not by their appearance but by the way they feel. Untreated mushrooms, if they are fresh, feel dry and smooth. The chemically treated mushrooms feel wet and slippery.

WILD MUSHROOMS

The chanterelle is shaped like a horn of plenty and is either gold or orange in color. It is one of the finest wild mushrooms. While it has been domesticated with some success, most of the ones that come to market are gathered wild in the evergreen forests of the Northwest. They are picked from the forest floor by highly skilled and highly paid mushroom hunters. The chanterelles are in season in late spring; they are gathered in the mountains as soon as the snow melts. They command very high prices.

The chanterelle has a jet-black cousin that is similar in size, shape, flavor, and texture. This variety is called the Black Trumpet and is rarer and costlier than the golden chanterelle.

The Shitaki is a very popular and very available variety. It originated in Japan but has been successfully cultivated in the United States. For obvious reasons the Shitaki has been renamed the Golden Oak mushroom. They grow on decayed oak logs. They are dark in color and have open caps that are very tender, but the stems are a bit chewy.

The Enoki is also a native of Japan that has found a home in the United States. Like the Shitaki, the Enoki spawns on decayed oak logs. They are tiny, fragile, and creamy white in color. The caps are the size of large pearls and are attached fairly long, very thin stems. The Enokis are most tender and are especially in demand in Oriental recipes.

The oyster mushroom is the easiest wild variety to cultivate and therefore is the least costly of the "wild" mushrooms. It is produced on an ever-increasing scale in Pennsylvania and on the West Coast. Oyster mushrooms are not highly regarded in Europe, but in the United States they are in great demand for use in Chinese restaurants. Oyster mushrooms are a light gray in color and somewhat resemble an oyster in shape. Like their namesake, they are highly perishable and have a short shelf life. Oyster mushrooms are fairly tough and chewy, much more so than other wild varieties. Their worst fault is that if they are even slightly overcooked, they tend to get sticky and gummy.

The Italian Brown, so named for want of a better one, is one of the more popular wild mushrooms and is imported from Italy. It is dark brown in color, quite large, and has open caps and long thick stems. They are very meaty yet very tender and have a rich, earthy flavor. If you have yet to cook wild mushrooms, I recommend this variety. They aren't quite as costly as some of the other wild varieties and they are quite similar to the more familiar white mushrooms.

The Portobello, or Roma, is another import from Italy. The caps resemble the shape and size of a pancake and the stems are long and thick. Some of these stems have a fair amount of black soil clinging to their bases that may be difficult to remove completely. Despite its not very attractive shape and appearance, the Portobello is a fine, tender, flavorful mushroom.

The Ceps, which are also known as Boletus, are the finest wild mushrooms. Only the morels and the truffles are more costly, and they aren't true mushrooms but a close relative. Unfortunately, like the morels and the truffles, the Ceps refuse to be domesticated. They are grown in Europe and have a short season, from June to November. However, dried Ceps may be purchased year-round.

The morel is the second most costly fungi. Even though they are alleged to be the most flavorful variety, they are possibly the least attractive, with an odd, spongelike appearance. The hollow caps are dark brown in color and conical in shape. The stems are lighter in color. They arrive in May and have a short season.

The truffle is the Rolls-Royce of the fungi family. Unlike most of the other edible, and inedible, fungi, truffles grow beneath the surface of the earth, making them most difficult to find. In Europe, dogs and hogs are trained to detect the faint but distinct garliclike odor emitted by the truffles and to sniff them out.

The most esteemed and highest priced of the truffles is the coal-black variety. It is known as the Black Truffle and is found mostly in France but also on a lesser scale in Spain, Italy, and Germany.

Great value is also placed on the White Truffle that grows in Italy. The Red Truffle is also a product of Europe, but it isn't as highly prized or priced as the Black or White Truffle. No American species of truffle is considered edible.

Squash

❦

PUMPKINS
·
SUMMER SQUASH
·
WINTER SQUASH

T he squash is an edible gourd and is of the same genus as melons and cucumbers. The varieties of squash come in a wide assortment of sizes (ranging from a few ounces to over one hundred pounds), shapes, and colors (white, yellow, orange, green, brown, gray, and light blue). These colors are usually solid, but some varieties are variegated and others have two solid colors on a single squash. There are countless varieties, too numerous to outline. However, squash may be divided into two distinct groups: those that have hard shells and those that have soft shells. The hard-shelled varieties are known as winter squash and are quite hardy—some have a shelf life that exceeds six months—and they usually have orange-colored flesh. The soft-shelled varieties are known as summer squash, are quite perishable, and have white flesh.

Back when we used blocks of ice as refrigeration, summer squash was available only in the late spring and summer months. They disappeared from the marketplace at first frost. Today, thanks to modern refrigeration and speedy transport, summer squash is in the market twelve months of the year. In Grandma's day winter squash was available in the fall and winter months and, depending on the variety, well into spring. But today winter squash is also in the market year-round. Summer varieties require refrigeration, but it is preferable not to refrigerate winter varieties if temperatures are below 70°F.

PUMPKINS

The pumpkin is the most colorful and shapely winter squash and comes in a wide variety of sizes. Some varieties that have been coddled by farmers competing in contests have reached weights of two hundred or even three hundred pounds. Pumpkins, along with Indian corn and ornamental gourds, are symbols of autumn.

Though it can be served as a cooked vegetable or used in making pies, the pumpkin is primarily used for decorative purposes. Ninety-nine percent of the fresh pumpkins sold in the marketplace are purchased in the two-to-three-week period prior to Halloween. The day after Halloween, the unsold pumpkins have about as much value as the unsold Christmas tree on Christmas day.

The traditional symmetrical-shaped, deep-orange-colored pumpkins used to make jack-o'-lanterns on Halloween are not the best variety to use for cooking or baking pies. They are far too stringy. A less colorful, squat, light brown variety called the cheese pumpkin is the best cooking variety. A small orange-skinned variety called the sugar pumpkin is also quite good.

Baking a pumpkin pie using fresh pumpkin is usually more trouble than it is worth. Using canned pumpkin requires a lot less effort, usually costs less, and the end product is as good as if not better than a pumpkin pie made from fresh pumpkin.

SUMMER SQUASH

Summer squash is widely used in all areas of the United States. The more important varieties include the green zucchini squash, the straight- and crook-neck yellow squash, and the flat, discus-shaped, white squash.

Summer squash is harvested prior to reaching full maturity and while the seeds are still tender enough to be edible. The flesh of the summer squash varieties is tender and string-free.

From colonial times until the mid-1930s, the yellow-skinned varieties of the summer squash were by far the most popular. The white summer squash came in a distant second. Green-skinned zucchini squash was a rarity and could only be found in retail stores that served Italian neighborhoods. Today, the green-skinned zucchini is by far the biggest seller. At New York's Hunts Point Market, 95 percent of the summer squash sold each year is zucchini.

However, there are still some regional preferences for yellow squash. The crook-neck yellow squash sells well in Dixie, as does the straight-neck yellow in the Midwest. The white squash, also called patty-pan, has all but disappeared, though of late it has been making a come-back as a gourmet item. There is a brand-new yellow zucchini that was introduced recently, but it is still too early to determine if it will be accepted by the public.

Zucchini is of Italian origin. It is called green squash in many areas, and in French restaurants and British cookbooks it is referred to as *courgette*. It is a favorite with many home gardeners because it thrives with very little care and has a high yield.

Unopened flower buds of zucchini are a gourmet item. When they are sautéed they are a flavor treat. These buds are very expensive when bought in fancy fruit shops, but are free for the gathering if you have zucchini growing in your backyard.

When shopping for summer squash, freshness is all-important. When fresh, it is sweet, but it can be bitter in flavor when aged. The firmness of the summer squash is the clue to its freshness. Soft squash is old squash. Check out both ends with gentle pressure. If it is soft and rubbery, don't buy it.

In summer squash, unlike winter squash but like cucumbers, the seeds as well as the rind are edible. Most of the flavor of the squash is in these seeds. The smaller the seed, the better the flavor and the smoother the texture. The smaller the squash, the smaller the seeds. Therefore, the smaller squash is more desirable. When shopping for zucchini and the straight-neck yellow variety, select squash that is at most about seven inches in length. Any squash that is larger may have tough seeds. All summer varieties are harvested prior to reaching full maturity. When they reach full maturity they become semihard-shelled squash and the seeds are too tough to consume.

Remember to select small, very firm squash for top quality and flavor. If the squash is soft or oversized, pass it by. Store it in the re-frigerator.

WINTER SQUASH

There are countless varieties of winter squash. While there are some regional preferences for other varieties, the top three best sellers nationwide
are acorn, butternut, and Hubbard. Most winter squash, if stored in a
cool, dry, well-ventilated area, will have a shelf life of three to four
months. (Most summer squash have a shelf life of less than a week.) In all
winter squash varieties the seeds are too large and tough to be edible.
Seeds of some varieties are dried, then roasted, and are edible if the outer
husk is removed.

Winter squash no longer plays an important role in American cuisine. However, back in Grandma's day, and especially in Great-grandma's
day, it was a big seller, because summer squash as well as most of the
other fresh vegetables were out of season during the winter and early
spring. In that bygone era, before butternut and acorn squash had yet to
arrive on the scene in any quantity, Hubbard squash played an all-important role. The Hubbard is a large, hard as a rock, winter variety that can
be either dark green, light blue, or orange in skin color. The orange-
colored flesh is flavorful and has a tender, string-free texture. It is usually
baked but can also be steamed or boiled. Its biggest shortcoming, and the
reason the Hubbard has gone out of vogue, is its large size. Hubbards
usually weigh from five to ten pounds—far too large for today's average
family. Prior to self-service, greengrocers would cut a chunk of Hubbard
squash to order at the time of purchase. Today's usually understaffed self-
service supermarkets can't be bothered; they prefer to handle the smaller-
sized acorn and butternut varieties.

The acorn is really a miniature version of the Hubbard squash. It
has a very hard-ridged rind, orange-colored flesh, and hard seeds that
resemble those found in a pumpkin. The outer skin is usually dark green
in color with a trace of small orange areas. Some varieties that are slate
blue in color have been phased out in deference to the public's preference
for the dark green color. Avoid acorn squash that are completely orange
in color; these are usually green ones that have matured and will be dry
and stringy in texture.

The ideal-sized acorn squash is large enough for two servings. Split
it in half, scoop out and discard the seeds, sprinkle some brown sugar in

the cavity, add a pat or two of butter, and bake it in the oven as you would a potato. Since acorn squash is sometimes slightly stringy, you may wish to remove it from the oven when it is soft. Mash the flesh with a fork and rebake it for a few minutes. Some recipes call for stuffing the acorn with ground meat or sausage. Baked acorn squash has a flavor and texture very similar to that of baked sweet potatoes.

Use extreme caution when splitting an acorn squash. They are difficult to cut and the knife may slip.

The butternut variety is a buff-colored winter squash. It is not quite as hard as the acorn or the Hubbard but is hard enough to be used as a deadly weapon. In shape it looks somewhat like a thick-necked bowling pin. Butternut may be baked in the half-shell like acorn squash, but it is usually cut into pieces, steamed or boiled, and then mashed. The butternut won't be as difficult or as dangerous to cut as the acorn. It is the easiest winter squash to prepare and is second to none in fine flavor and smooth, string-free texture.

Both acorn and butternut squash are available year-round, but those sold in the fall and winter are usually of better quality than those that are in market in the spring and summer.

Other Squash

BANANA SQUASH
·
BUTTERCUP SQUASH
·
CHAYOTE
·
DELICATA SQUASH
·
GOLD NUGGET
·
KABACHI
·
MARROW SQUASH
·
SPAGHETTI SQUASH
·
SWEET DUMPLING
·
TURBAN SQUASH

I n the past ten years, as more unusual fruits and vegetables have increased in popularity, there has been an influx of less familiar-looking hard-shelled and semihard-shelled varieties of squash in the marketplace. Despite the fact that they are quite costly, they seem to sell fairly well. Most of these unusual squash are grown in California.

Credit for this deluge as well as for other little known fruits and vegetables goes to Frieda Caplan of Los Angeles, California, who heads a firm that specializes in promoting and marketing unusual fresh produce. She has appeared on countless TV and radio shows and has been highlighted in magazine articles. The products her firm sells are accompanied by recipes and cooking instructions. The following is a list and description of the off-beat squash that may be available in your area. All these squash are either hard- or semihard-shelled and nearly all may be baked, steamed, or boiled; they may also be cooked in microwave ovens.

Chayotes are available twelve months a year and require refrigeration. Supplies of other squash are erratic and limited; these other squash should be refrigerated if temperatures exceed 70°F.

Except perhaps for Marrow squash, as far as flavor and texture are concerned, none of these varieties is superior to, and some aren't nearly as good as, the more readily available, far less costly butternut squash. However, as they gain popularity and become better known, supplies will increase, more stores will handle them, and prices will ease.

BANANA SQUASH

This variety attains good size and, like most of the other hard- and semi-hard-shelled varieties, has an orange flesh. The skin color is usually a light orange, but some varieties are almost ivory-colored. The banana squash has good flavor and texture. While this variety is a newcomer to most areas east of the Mississippi, it is fairly well known on the West Coast.

BUTTERCUP SQUASH

This is a medium-sized variety. It is fairly round in shape and has a somewhat small bump at the blossom end. It is dark green in color with light green stripes. It has good flavor but is slightly dryer than most of the other hard-shelled squash.

CHAYOTE

This unusual squash is described on page 231, in the chapter on tropical vegetables.

DELICATA SQUASH

This variety is also called Bohemian squash. It is long in shape and the skin color is green with light brown stripes. It has a yellow flesh that is fairly sweet in flavor.

GOLD NUGGET

This is a small, round, orange-skinned squash that looks like a miniature pumpkin. This variety tastes best when baked on the half-shell, like acorn squash.

KABACHI

As you can guess by the name, this is an import from Japan, but it is also grown in California. It is pale green in color and round but flat in shape, and it has orange flesh. This variety may be found in Oriental food stores as well as in specialty fruit shops.

MARROW SQUASH

This is a popular variety in Britain, where it is known as English Marrow. In the United States it is seldom sold in retail stores but is popular with home gardeners. According to gourmets, the English Marrow is the best of the hard-shelled varieties of squash.

SPAGHETTI SQUASH

This variety has been around for longer than most of the other unusual squash varieties. It made a big hit when it was first introduced in the 1970s, but sales are now well below that peak and slipping. This is a semihard-shelled squash. It has a bright-yellow-colored skin and a very stringy white flesh. After you boil this variety, the flesh can easily be removed with a fork. These strands of squash can be used instead of pasta, served with a tomato sauce. The end result is a pasta-type dish that is made from a fresh vegetable rather than semolina flour. It is a lot of fun and a conversation piece the first few times it is served, but since the preparation is somewhat time consuming, the novelty may soon wear thin.

SWEET DUMPLING

This variety is small, round, and green-and-white-striped in color. It has the traditional orange flesh of winter squash and is fairly sweet with a smooth texture.

TURBAN SQUASH

This appropriately named variety resembles a colorful turban. It is usually orange in color at the base (the stem end) and has a small green knob at the top. There is no doubt that the turban squash makes a colorful table decoration, but there is some doubt as to its eating quality.

One-of-a-Kind Vegetables

❧

ASPARAGUS
·
CARDOON
·
GLOBE ARTICHOKES
·
JERUSALEM ARTICHOKES (SUN CHOKES)
·
OKRA
·
RHUBARB
·
SWEET CORN
·
SWEET POTATOES

T he vegetables in the preceding chapters were grouped together because they belong to the same botanical family or have common characteristics. This chapter covers the one-of-a-kind vegetables that are akin to no other edible varieties.

ASPARAGUS

Asparagus is the aristocrat of the vegetable kingdom. A member of the lily family, it is believed to have originated in Asia Minor. The ancient Greeks and Romans prized it highly as both a food and as a medicine.

Back when we still had trolley cars, the arrival of fresh asparagus was the first herald of spring. It made its annual debut late in February and bowed out of season in mid-July. Today, thanks to improved agricultural know-how, modern refrigeration, and especially the speed of the cargo jet, this former rite of spring has been extended to cover all four seasons and is now available year-round. In January we get asparagus from Mexico, Chile, and New Zealand. In February the California crop comes on line and is available until July, at which time we get asparagus from Washington and Michigan. Late in September we get a second crop from Mexico. In October, November, and December we again fly in asparagus from Chile and New Zealand.

California is by far our best and largest source of "grass" (a market term for asparagus long before it became popular in reference to marijuana). California is also a major exporter of asparagus, via jet, to Europe. The California season reaches peak in mid-April and, weather permitting, it is sustained until the end of June, at which time Washington State and Michigan take up the slack.

Fresh asparagus is also produced on a more limited scale in Oregon, Pennsylvania, and New Jersey, but these crops are mostly sold locally. Not too long ago New Jersey was a major source of supply. However, in recent years their tonnage has significantly decreased due to their crops being ravaged by a plant disease known as fusarium root rot.

When shopping for asparagus, look for the two hallmarks that identify quality: the condition of the tips and the color of the spears. Perfect asparagus will have tips that are dry, tight, firm, and purplish in hue, and come to a point. If the tips are feathery, open, and soft, or have gone to seed, that asparagus is past its prime but still usable. However, if the tips are wet or slimy, they are on the verge of, or have started to, decay and should be avoided. Aged or decayed asparagus emits a most unpleasant odor. If you have any doubts about its freshness, your nose is your best guide.

There are some white (actually oyster-colored) varieties of asparagus. They have a stronger, more earthy, taste than the green types. While this white asparagus is highly prized in Europe, it is seldom sold in the United States. We prefer the more tender green varieties.

When purchasing asparagus, the price tag is not a true barometer of its value. In the green varieties, the greener the spear, the better the quality and the greater the yield. Even in all-green asparagus, at least 15 percent to 20 percent of the spear is too woody and chewy to be palatable. In spears that are half white and half green, the bottom 70 percent of the spear is inedible. Spears that are green only near the tip can have as much as 90 percent waste. Therefore, all-green asparagus is always a better buy than asparagus that is less green selling for half the price. Since asparagus is usually quite costly, the prudent shopper should always check out the color of the spears as well as the condition of the tips prior to making a purchase.

The presence of sand or loam in fresh asparagus can be a gritty problem. When sand gets lodged in the tips, no amount of rinsing with water will totally remove it. If there is any trace of sand in the asparagus you find in your market, pick up a bunch of broccoli in its stead. Asparagus from New Jersey is more likely to have sand than asparagus from other areas. California asparagus rarely has this problem.

Asparagus spears can be either pencil thin or as thick as your thumb. In asparagus of comparable quality (condition of the tips and color of the spears), the thickness has little bearing on the quality. However, as a rule, the thicker the spear, the higher the price. The *haute cuisine* eateries will pay premium prices for the thickest asparagus, yet in Italian neighborhoods merchants can't give thick asparagus away. There, the very thinnest spears (in the trade they are called spaghetti) are most highly prized. A traditional Neapolitan recipe calls for frying very thin asparagus with eggs.

Until about twenty years ago, all the asparagus that came to market was carefully graded by thickness and packaged in 2¾ pound bunches. In order of decreasing thickness, the spears were graded as colossal, extra select, select, extra fancy, fancy, and standard. Today asparagus comes loosely packed in fifteen- or thirty-pound boxes. It is no longer bunched or accurately graded by size or thickness. When you purchase asparagus, try to select spears that are uniform in thickness. If there is too great a range in size, the thinner spears will be overcooked by the time the thicker ones are just tender.

To get the most mileage out of asparagus, here is an old wives' tale that really works: Never cut the spears with a knife. Snap them as you would a twig or a pencil. This is easily accomplished because fresh asparagus is quite brittle. Magically, where the spear snaps is the dividing point between the tender and tough portions. When you use this method your end product may not be uniform in length, but it will be uniform in texture and tenderness and you won't be discarding any edible asparagus. If the asparagus is fresh, there is no need to peel or pare it with a knife. As with fresh mushrooms, peeling the spears is not only a waste of time but also a waste of asparagus.

□ WHEN TO BUY: *Available year-round; At peak: March to July*
□ WHAT TO LOOK FOR: *Dry, tight, firm, purplish tips that come to a point and vegetables have a fresh odor*
□ HOW TO STORE: *Refrigerate immediately after purchase*

CARDOON

Cardoon, also called cardoni, is native to southern Europe and North Africa. Although it is related to the artichoke, it looks more like a very large, slightly thorned, coarse bunch of celery. It is often two feet in length and isn't very attractive. However, it is highly prized by southern Italians and Sicilians.

The limited amount of cardoon grown in the United States is produced in California. The peak of season is in the fall and winter months. The coarse outer ribs are tough and stringy and are usually discarded. The inner stalks and the heart are tender when boiled (they are usually fried after they are boiled). Like its cousin the artichoke, the cardoon tends to discolor and blacken when cooked in aluminum utensils. Vinegar or lemon juice will retard or eliminate this discoloration.

☐ WHEN TO BUY: *Supplies limited and erratic; available late fall through winter*
☐ WHAT TO LOOK FOR: *Firm, erect stalks*
☐ HOW TO STORE: *Requires refrigeration*

GLOBE ARTICHOKES

The artichoke is an unusual-looking vegetable that somewhat resembles an oversized olive-green pine cone with a sharp barb on the end of each leaf. The globe artichoke is the unopened flower bud of a plant that belongs to the same genus as the thistle. Despite its rather exotic appearance it has a mild, pleasant, nutty flavor.

Nearly all the globe artichokes produced in the United States are grown in an area along the California coast just below San Francisco. The small town of Castroville, California, claims the title "Artichoke Capital of the World." This claim is grandiose because by far the world's number one artichoke producer and consumer is Italy, followed by Spain, France, and Greece. While the United States is an also-ran in volume of tonnage, it is second to none in quality of the product.

Artichokes thrive in a climate that is neither too warm nor too sunny but one that is moist and damp. That describes the climate of the area around Castroville. Each morning for ten months of the year, the same thick, clammy fog that plagues nearby San Francisco rolls in from the Pacific and shelters the artichoke plants from direct sunlight.

Although artichokes are available year-round, the peak of season covers the months of March, April, and May, during which time more than 50 percent of the annual crop is shipped to market. The poorest times of the year are in July and August, when the weather in the artichoke-growing area is hot, dry, and almost fog-free. These summer artichokes, which usually have a purplish hue along with the normal olive-green coloring, are less tender than those produced in the peak of season.

While artichokes dislike hot weather, they also are affected by cold weather. When exposed to a heavy frost they die, but a light frost affects their outer appearance without affecting the eating quality. When exposed to frost, the skin of the artichoke starts to blister and peel (not unlike sunburn on humans). The olive-green color turns to bronze. The growers claim that these "winter-kissed artichokes" are superior in flavor to the more attractive ones. This claim is strictly PR. However, while the discolored artichokes are not more flavorful, neither are they less flavorful. As a rule, these winter-kissed artichokes are offered at modest prices. If they are firm and tight and if they look fresh and green when you peek in at their inner leaves and at the base, ignore the discoloration and enjoy the good flavor and low price. However, if the heart of the artichoke isn't a light green in color, or if it's soft, flabby, and the leaves have opened up, pass it by. Note that discolored artichokes are only a good buy during the winter months. During the rest of the year, a discolored, bronze, or brown artichoke is not winter kissed but probably too old.

Selecting top-quality artichokes is quite easy. When they are not fresh they will look dull, old, and tired and will be soft, open-leaved, and discolored. When they are fresh they are firm, compact, and a very attractive olive-green in color. The other hallmarks of quality are the shape, firmness, and weight in relation to size. The perfect artichoke is round as a baseball. A blunt, round artichoke has more leaves and a bigger heart and is heavier than one that is angular and comes to a point. As a rule, the artichokes produced in the winter and spring are rounder than those grown in the summer and fall.

Artichokes come in various sizes, ranging from as small as a plum to almost as big as a softball, including all sizes in between. However, the size has no bearing on the quality and flavor. A fine small artichoke is every bit as good as a very large one of comparable quality. In Italy, and in Italian neighborhoods in this country, the small artichokes are highly prized. In France the very large ones are preferred.

Most cookbooks give detailed instructions on how to prepare and cook artichokes. Remember that they tend to discolor in the cooking process. Adding a few drops of vinegar or fresh lemon juice to the water before cooking will retard discoloration. If you want to be a purist, use a stainless steel knife to trim and cut the artichoke, since a carbon steel knife may discolor it.

In the winter months, if the prices of the California artichokes reach very high levels, some are flown in from Chile. These imports, while selling for top dollar, are not always of top quality.

Since artichokes are quite perishable, use them as soon as possible after purchasing them. However, they will hold up for at least a week without breaking down if stored in the refrigerator.

☐ WHEN TO BUY: *Available year-round but at peak in March, April, and May*
☐ WHAT TO LOOK FOR: *Firm, compact vegetables with attractive olive-green coloring that are heavy in relation to their size*
☐ HOW TO STORE: *Refrigerate immediately after purchase*

JERUSALEM ARTICHOKES
(SUN CHOKES)

Jerusalem artichokes, native to North America, were used by the Indians long before the coming of the settlers from Europe. Their name is a misnomer on both counts. They have no connection with Jerusalem and they are not artichokes, which is why they have been renamed and are now called sun chokes. They are tubers produced beneath the ground by a variety of sunflower. They look like small, knobby, white potatoes or a piece of ginger root.

Sun chokes are grown in California on a limited scale that is more than adequate to satisfy the limited demand and are therefore sold primarily in the more costly gourmet produce outlets. They are first harvested in the fall and continue to be available for most of the winter.

Jerusalem artichokes may be eaten raw, boiled, or sautéed. They tend to discolor and blacken if cooked in aluminum pots or pans. They are pure white in flesh and have an unusually earthy but pleasant flavor when served raw in salads.

Choose firm, light-colored sun chokes (they are usually sold prepackaged) and store in the refrigerator. Look for recipes on the package.

☐ WHEN TO BUY: *Supplies erratic; available late Fall through Winter*
☐ WHAT TO LOOK FOR: *Firm, light-colored vegetables*
☐ HOW TO STORE: *Require refrigeration*

OKRA

Okra is believed to have originated in that part of northeastern Africa that we now know as Ethiopa and is now a staple in much of Africa, the Near East, and the countries at the eastern end of the Mediterranean Sea.

Okra arrived in our southern states in the holds of slave ships. It was introduced in Louisiana by the French colonists and in the rest of North America as an ingredient in Campbell's Chicken Gumbo Soup.

Although okra looks somewhat like a legume, it is a member of the Hibiscus family and is closely related to cotton. It thrives in areas that have long, hot, not too wet summers. In the United States we grow okra in Florida, Texas, Georgia, and New Jersey. In midwinter some okra is imported from Mexico.

When properly prepared, okra is a flavorful vegetable, but when it is overcooked, the pods open up and attain a mucilaginous consistency and the result is a gluey mess.

Okra may be boiled, baked, or fried, but it is used mainly in Creole cooking. In Dixie, fried okra is a prized delicacy, and in New Orleans gumbo recipes, okra is combined with chicken or shellfish.

Okra, like artichokes, tends to discolor in the cooking process, especially when cooked in aluminum pots. It causes a harmless chemical reaction that discolors the utensils as well as darkens the okra. The trick to retarding and even eliminating this discoloration is to add fresh lemon juice or vinegar to the water. This also keeps the okra pods from opening up.

Just as special care is needed when cooking okra, special care is also a must when shopping for it. You must be selective. Top-quality okra is small and feels like velvet to the touch. The smaller the pod, the more tender the okra. Pods under two and a half inches in length are ideal. The larger pods, especially those over four inches, are usually tough, woody, and too fibrous to chew.

Okra can also be found in the frozen-foods section. The baby whole frozen okra is far superior to the frozen cut okra.

- □ WHEN TO BUY: *Available year-round*
- □ WHAT TO LOOK FOR: *Small pods, under two and a half inches in length, that are velvety to the touch*
- □ HOW TO STORE: *Requires refrigeration*

RHUBARB

In ancient China rhubarb was used for medicinal purposes, and even as recently as our colonial times it was believed to cure tired blood.

Rhubarb wasn't recognized as a food until about three hundred years ago. Just as the tomato is a fruit but is referred to as a vegetable, the rhubarb is a vegetable but is thought of as a fruit. It is used in pies and sauces and is especially good when cooked with strawberries. In some areas of the United States rhubarb is known as the pieplant.

There are two types of Rhubarb: outdoor and hothouse. Outdoor rhubarb comes in season in spring and lasts until fall. The stalks are often more green than red and have large, floppy, green leaves. It is quite coarse in texture and often stringy. It is also very tart and requires lots of sugar. The hothouse variety, which is produced in California, Oregon, and Michigan arrives in January and winds up in June. It is either cherry-red or blushing-pink. Some hothouse rhubarb have bright yellow leaves and are particularly attractive. It is not as stringy as outdoor rhubarb, has a milder flavor, and requires less sugar. Rhubarb leaves, which are toxic, should never be eaten.

Fresh rhubarb can be easily identified by its firm and erect stalks. When it's not fresh, the stalks are limp and floppy.

☐ WHEN TO BUY: *January through August*
☐ WHAT TO LOOK FOR: *Firm, erect stalks*
☐ HOW TO STORE: *Refrigerate immediately after purchase*

SWEET CORN

Corn, which is native to America, is botanically related to wheat, barley, and rye. Its proper name is *maize,* and that is what it is called in most of Europe. The term *corn* is used in the Old Testament but refers to all the other grains and cereals that were known in that period. Corn or maize as we know it didn't reach Europe until the sixteenth century.

There are two types of corn: field corn and sweet corn. Field corn is dried on the cob, then shucked and sold as grain. It is used to feed cattle and hogs. It is also ground into meal (flour), and pressed to yield oil. Sweet corn is harvested while it is young and tender and is used as a fresh vegetable. It is also frozen or packed in tins.

While sweet corn on the cob is probably one of the best-loved vegetables in North America, it isn't as popular in Europe or the rest of the world. The flavor of no other fresh vegetable is as perishable as that of corn. The speed with which it is transported from the field to the pot of boiling water determines the flavor and the texture of the ear of corn. Corn on the cob, like fresh lobster, has to be cooked while it is still alive.

Corn is at its tastiest, sweetest, juiciest, and tenderest the instant it is severed from the stalk in the cornfield. At the moment of separation the ear starts to heat up and dehydrate and the sugar in the kernels starts to change into starch. After about forty-eight hours of heat and dehydration, the once sweet and juicy kernels become dry and starchy. This chemical change is relentless; however, it can be checked, but not halted, by precooling and constant refrigeration. In rural areas, where every home has some sweet corn growing in its backyard, it is claimed that first the water in the pot is brought to a boil and then the corn is picked.

Corn can have either yellow or white kernels or a combination of the two. But the color is not a clue to the flavor. Only the freshness of the ear determines that flavor. Whether it was grown on a plush estate or on the wrong side of the tracks, the bottom line is how long it took to get the corn from the field to the kitchen.

Corn that is used the day it is picked will be superb. If it is a few days old and has been precooled and stored under constant refrigeration, it will still be quite good. This explains why some, not all, of the corn grown in Florida and California in midwinter and shipped north is of acceptable quality. However, if the corn has been neglected and allowed to heat up and dehydrate, even if it is at the peak of season, the corn will be hard and dry.

When shopping for corn, look for green (not brown or discolored) outer husks and plump, firm kernels. If the kernels have dimples, the corn is too old to enjoy and will be tough and dry. Never purchase corn if the ear feels hot to the touch. Even corn that is picked in 90-degree weather will feel cool to the touch. Once the ear has started to heat up, it cannot be checked, even if refrigerated.

The ability to purchase freshly picked corn will depend on your geographical location. If you live in a big city, your chances are almost nil—at best you'll be able to purchase only fair corn. If you live in the suburbs, you'll have a fighting chance. Most suburbs have road stands that specialize in freshly picked corn. Unfortunately, most suburban supermarkets are supplied by their own warehouses and offer the same poor quality corn as the big city markets.

☐ WHEN TO BUY: *Available year-round depending on geographical location*
☐ WHAT TO LOOK FOR: *Freshly picked ears with plump kernels that are cool to the touch*
☐ HOW TO STORE: *Refrigerate immediately after purchase*

SWEET POTATOES

The sweet potato is a member of the morning glory family. It originated in Central America but is now grown in all areas of the world that have subtropical climates. It is a near-perfect food and is a staple in many of the world's underdeveloped nations.

We mistakenly call one type of sweet potato a yam. The yam is of a different botanical genus, but is often used as a substitute for the sweet potato.

There are two types of sweet potatoes. The ones that we often mistakenly call yams have a deep orange-colored flesh and are very moist and sweet when cooked. They are grown in most of the southern states, with North Carolina, California, and Louisiana as the top producers. Thanks to a curing process, these sweet potatoes are available twelve months of the year.

The other type of sweet potato, which is known as the white sweet or Jersey sweet, is pale yellow or off-white in flesh color. These are not nearly as moist and sweet as the orange-fleshed variety and, except for limited pockets of regional preference (Philadelphia and South Jersey), have been almost completely displaced. Yet in Grandma's day the white sweets had about 90 percent of the market.

While sweet potatoes are available year-round, the two lightest months are June and July. During those months they are nearly a year old and the quality slips as the price rises. The most flavorful sweet potatoes arrive in market during August, September, and October. These are uncured and shipped to market directly from the field.

Freshly harvested sweet potatoes have a very short shelf life. Because they have a very high moisture content, they are prone to spot and decay. At best they last four to six weeks before they start to break down. By placing them in a kiln and removing much of this moisture, their shelf life can be extended from a few weeks to more than eight months. Kiln-dried sweet potatoes are called cured; those that are not kiln-dried are called uncured.

The cured sweet potato isn't quite as soft and moist as the uncured sweet potato; so those that are purchased from August through October will have the best flavor. All the sweet potatoes that are in market during that three-month period are uncured; those sold from November until the following August, when the new crop arrives, are cured. There is only one crop per year.

Sweet potatoes are a subtropical vegetable that thrive in warm weather and can't tolerate the cold. Never store them in the refrigerator or they will cut black after cooking. Sweet potatoes are almost as sensitive to refrigeration as bananas are.

When selecting sweet potatoes, always choose those that are firm and shapely and have clear, unmarked skins. Those of medium size are your best bet if you are going to bake them whole, but they usually sell at premium prices. Very small ones taste fine, but may be more trouble to prepare than they are worth. Jumbo ones are good for boiling but take a long time to bake. However, the jumbos usually sell for about half the price of medium sweet potatoes, at least at the wholesale level.

☐ WHEN TO BUY: *Available year-round but at peak in August, September, and October*
☐ WHAT TO LOOK FOR: *Firm, shapely vegetables of medium size with clear, unmarked skins*
☐ HOW TO STORE: *Do not refrigerate*

Oriental Vegetables

LONG BEANS

·

BOK CHOY

·

CHINESE CABBAGE

·

DAIKON

·

GINGERROOT

·

CHINESE MUSTARD GREENS

·

BITTER MELON

·

WINTER MELON

NAPPA

·

CHINESE OKRA

·

SNOW PEAS

·

LOTUS ROOTS

·

BEAN SPROUTS

·

CHINESE SQUASH

·

TARO

·

TOFU

U ntil recently, few Americans had access to the vegetables used in Chinese cuisine. Today, with Chinese and Oriental cooking very much in vogue, Oriental vegetables are available in most of our larger supermarkets. Snow peas, bean sprouts, tofu, fresh gingerroot, daikon, bok choy, and Chinese cabbage sell well and are fairly easy to find. Some of the more unusual items may be available only in larger cities or Oriental neighborhoods.

LONG BEANS

Long beans (dow kok) are pencil thin, tender green beans that are about a foot long and very much resemble the French haricots verts. They are usually stir-fried.

BOK CHOY

Bok choy is an Oriental version of Swiss chard. It not only can be used when called for in Oriental recipes but also as a substitute for fresh spinach.

CHINESE CABBAGE

Chinese cabbage is a long, thin, not quite as compact version of nappa.

DAIKON

Daikon (low bok) is a large white radish that is shaped like, but is at least twice as thick as, a large carrot. Despite its size it is crisp and juicy and has a nice, not too strong flavor. It can be used either raw or cooked. Pickled daikon is highly prized in Japan, where no rice dish is ever served without it. In tonnage the daikon is Japan's number one vegetable crop.

GINGERROOT

Fresh gingerroot is an indispensable flavoring ingredient in a host of Oriental recipes. The flavor of fresh ginger is far superior to the ground ginger powder that you find on the supermarket spice rack. It looks somewhat like an elongated, very gnarled, thin-skinned new white potato. Fresh gingerroot is in the market year-round. We import it from as far away as the Fiji Islands, but the finest fresh ginger is grown in Hawaii. It has a fairly long shelf life and a small amount goes a long way. It lasts for weeks if refrigerated, but if you slice gingerroot and store it in sherry wine, it will last for months.

CHINESE MUSTARD GREENS

Chinese mustard greens (kai choy) have white stems and green leaves. They are mildly bitter in flavor and are either fried or used in soups.

BITTER MELON

A bitter melon (foo gwa) looks like a cucumber with ridged, bumpy striations. The inner flesh is white and pink with seeds that resemble those of cucumbers. It is primarily used in soups but may also be fried.

WINTER MELON

A winter melon (dong qua) looks like a small round watermelon. The pale green skin usually has a white waxlike coating. It has a solid white flesh that contains seeds. The winter melons are used in soups and are especially highly prized when used in shrimp dishes.

NAPPA

Nappa is a squat, compact, very tender cabbage. It can be used either as a cooked vegetable or raw in salads.

CHINESE OKRA

Chinese okra (sing gwa) are about a foot long and about an inch in diameter. The skin is dark green and deeply ridged. Chinese okra is usually used in seafood dishes.

SNOW PEAS

Snow peas (hon-lon-dow), the most popular Oriental vegetables, are discussed with other legumes on page 166.

LOTUS ROOTS

Lotus roots (lin gow) are the roots of a variety of lily. They are a favorite in all types of Chinese cuisines but are primarily used in soups.

BEAN SPROUTS

Bean sprouts are a nutritious yet economical vegetable and are equally flavorful when served either raw or cooked. They are made by soaking dried soybeans or mung beans and allowing them to sprout. The mung bean sprouts are more popular. They are smaller and more tender than the soybean sprouts.

CHINESE SQUASH

Chinese squash (mo qua) have a fuzzy light green skin and are about five to six inches long. They are usually stuffed with rice and meat or seafood, used in soups, or stir-fried.

TARO

Taro (dasheen) is the tropical counterpart to our white potato. While it is starchy, it is very digestible and is often used in the diets of infants and invalids. Like the potato, it may be boiled, mashed, fried, or baked. Taro is a staple in the diets of native Hawaiians who use it to make poi.

TOFU

Tofu is a bean curd. It is made by grinding up soybeans and allowing them to coagulate. It is then pressed to remove moisture and cut into squares. The end product looks like small blocks of semisoft white cheese. According to some nutritionists tofu is the perfect food. It is very high in protein and is often used as a meat substitute by vegetarians. Tofu is semiperishable; store it in the refrigerator as you would cheese.

Tropical Vegetables

CALABAZO
·
CHAYOTE
·
JICAMA
·
NOPALES
·
PLANTAINS
·
YAMMI
·
YAUTIA
·
YUCCA

The tropical vegetables used to be available only in areas that had large Hispanic populations. Today, thanks to the increased popularity of Mexican cuisine and Mexican restaurants, tropical vegetables are starting to appear, although in limited supply, in the larger supermarkets in areas with minimal Latin American populations.

Words like *taco*, *tortilla*, *tamale*, *enchilada*, *chili con carne*, *guacamole*, and *gazpacho* have entered our vocabulary. It is only a matter of time before the names of the tropical vegetables enjoyed by Latin Americans will also have common usage. Words like *calabazo*, *chayote*, *jicama*, *nopales*, *plaintain*, *yammi*, *yautia*, and *yucca* are appearing in recipes in magazines and newspapers. Just as avocados, mangoes, kiwis, and papayas were sold only as curios fifty years ago and today are available in ample supply in most food markets, tropical vegetables may soon enjoy the same consumer acceptance.

All tropical vegetables are available twelve months of the year depending on geographical location and demand.

CALABAZO

The calabazo is a large, hard-shelled gourd that looks like a cross between a pumpkin and a Hubbard squash. It has a smooth skin and a thick, orange-colored flesh. It may be baked or boiled in the same manner as winter squash. The calabazos are usually too large for an average family to handle and are cut into sections by the retail markets. Uncut calabazos do not require refrigeration unless temperatures exceed 75°F.

CHAYOTE

The chayote is an unusual-looking but fine tasting squash. It is pear-shaped, resembles a pale green avocado, and is often covered with hair-like spines that are usually removed prior to shipment to market. The chayote has a single flat seed.

The flesh is fiber-free and similar in texture to zucchini. Unlike most other varieties of squash, the chayote retains much of its crisp texture after being cooked. It may be cooked with other vegetables, but is flavorful enough to stand alone. The natural seed cavity makes it ideal to use as a boat. Stuff it with cheese or ground beef and then serve it on the half-shell after baking it. One-half of a chayote makes an ample but low-calorie serving. Chayotes require refrigeration.

JICAMA

The jicama is a white-fleshed tuberous root vegetable that plays the same role in the tropics as the white (Irish) potatoes in North America. The jicama is a member of the morning glory family and looks like a large yellow turnip. They flourish in tropical climes—in areas that are too warm for growing potatoes.

It has a crisp, crunchy flesh and a rather bland flavor that has just a hint of sweetness. Unlike the white potato, the jicama is excellent when eaten raw, especially when sprinkled with a few drops of fresh lime juice. It may be used raw in salads or cut into sticks and served with dips. It is often served with guacamole. It may be boiled, mashed, fried, or baked, as you would prepare white potatoes. Even after it is cooked, the jicama retains much of its crunchy, crisp texture. Jicamas are often substituted for the more costly, equally crisp, water chestnuts. They should be refrigerated after purchase.

NOPALES

The nopales are the large round leaves of a variety of cactus. They are the size and the shape of a dinner plate, a light green in color, and about three quarters of an inch thick. When harvested they are covered with sharp barbs, most of which are removed prior to shipment to market. But handle them gingerly, because the barbs are usually not completely removed. Nopales should be refrigerated after purchase.

Nopales take little time to prepare. The leaves are cut into strips and then have to be peeled. The strips are then sliced as you would French a string bean. They may be boiled or fried. Use them as you would green beans, which they taste like but are crisper in texture than.

PLANTAINS

Plantains look like overgrown green bananas. They are an important food staple in most tropical nations. Unlike their cousins the bananas, plantains are too starchy to eat raw. They are prepared like potatoes—either boiled, mashed, baked, or fried. When cut into wafer-thin slices and then fried, plantains are more flavorful and nutritious than potato chips. Plantains need refrigeration only if temperatures exceed 75°F.

YAMMI

The yammi is a root vegetable and is the true yam. Like the yam, it does not require refrigeration. It has a white flesh and a sweet flavor. It is the counterpart to our sweet potato and may be boiled, baked, or fried.

YAUTIA

Yautia is a tuberous root and is the counterpart to our white potato.

YUCCA

Yucca is also a root vegetable and is known as cassava. It may be boiled or baked. It is also used to make tapioca.

Fresh Herbs

BASIL
·
CHIVES
·
CORIANDER
·
DILL WEED
·
MARJORAM
·
MINT
·
OREGANO
·
PARSLEY
·
ROSEMARY
·
SAGE
·
TARRAGON
·
THYME

I n botany an herb is any vegetation that doesn't have woody tissue and that dies after it has borne fruit. The Latin word for grass is *herba*. In culinary lexicon an herb refers to a special group of aromatic, flavorful plants that are added to complement the flavors of other foods.

Until quite recently most fresh herbs were available only to home gardeners during the growing season. Today fresh herbs may be found in the larger supermarkets and some specialty produce stores. Although most herbs are available year-round, supplies are often erratic.

During the off seasons fresh herbs are grown in hothouses or brought in via air. They usually carry high price tags, but since the addition of a few sprigs of an herb can add zest and sparkle to a bland recipe, fresh herbs are worth the high cost. In addition, they add flavor to foods without adding calories.

Fresh herbs are sold cut and banded in small bunches and at most have a shelf life of only a few days. Then they start to wither and lose color. Purchase only those that look fresh and alive; pass up those that are limp or have started to yellow. Try not to purchase herbs more than a day or two prior to use and keep them in the refrigerator.

Even though the bunches are small, you'll probably have more herb than called for in your recipe. At best, they will only last a week or so under refrigeration. However, the excess herbs may be salvaged by freezing or drying.

Remember to use herbs sparingly; overuse can overpower and drown out the more subtle flavors of the other ingredients in a recipe.

BASIL

Basil has a sweet flavor and is most aromatic. Since it blends well with garlic and raw or cooked tomatoes, it is frequently called for in Italian recipes. It is a basic ingredient in pesto. Cut basil is offered in small bunches year-round. During the summer months it is also available in pots. While most basil is green in color, there is also a reddish-purple variety. Both types are equally fragrant and flavorful.

CHIVES

Chives, which look like lush green grass, are delicate green onions. Bunches of cut chives are available twelve months of the year; in the spring they are also available in pots. Chives have a moderate onion flavor and a bright green color. They make an interesting and colorful addition to dairy products and omelettes.

CORIANDER

Coriander is also known as Chinese parsley, and in Latin-American neighborhoods it is called cilantro. It is a major seasoning in Mexican and Oriental cooking, and is the strongest flavored and most aromatic parsley. It has flat leaves that are not quite as crisp or colorful as the other types of parsley, but it should not be purchased if the leaves have started to yellow.

DILL WEED

Fresh dill has delicate, feathery, bright green leaves. It has a savory, sharp flavor and a fragrant aroma. Small bunches of fresh dill are readily available in most markets and are in season year-round. Dill is used in sauces and salads. It adds zest to chicken soup and a few sprigs added to potato salad livens up the flavor.

MARJORAM

Marjoram is a member of the mint family. It has grayish-green ovate leaves that are fragrant. The flavor resembles that of oregano. Since marjoram and oregano are similar in flavor and aroma, one may be substituted for the other.

MINT

Mint is a very common herb and is by far the easiest to grow. It is at flavor best when used in cold drinks during the summer. Fresh mint or a mint sauce or jelly are most compatible with roast lamb. Dry mint leaves add an unusual flavor and fragrance to tea.

OREGANO

Oregano, which is related to mint and marjoram, is a frequently used seasoning in Italian cuisine. Its pungent flavor and aroma spark tomato dishes, especially a tomato based spaghetti sauce. While it isn't quite as green in color as most of the other fresh herbs, oregano has a fairly long shelf life.

PARSLEY

Parsley is our most common herb. There are two types. The curly variety is prettier and greener than the flat-leafed type. The curly-leafed parsley is mainly used as a garnish. The less attractive flat-leafed variety, which is also known as Italian parsley, has a stronger flavor and is primarily used in cooking.

ROSEMARY

Rosemary is both spicy and fragrant. While it is primarily used to flavor fish and meat dishes, it is also used to enhance the flavor of the more bland vegetables and is especially good when added to boiled potatoes.

SAGE

Sage features gray-green furry leaves that have a sharp, musty flavor. It is used to flavor meat and poultry dishes and is a must ingredient in making pork sausage.

TARRAGON

Tarragon has slender green leaves that are aromatic and have a slight aniselike flavor. It is widely used in French cuisine and is often used to flavor wine vinegars.

THYME

Thyme has aromatic leaves that are pungent and spicy. It is used to flavor fish chowders and is an important ingredient in Creole cooking and in poultry seasoning.

METHODS OF PRESERVING HERBS

Freezing ❧

Most herbs freeze well with little or no loss of aroma or flavor. Wash the herbs and pat them dry. Those with fairly large leaves should be stripped and the stems discarded. Those with small leaves should be frozen, stem and all.

Place the herbs in a plastic container. (A container is preferable to a plastic bag because frozen herbs are brittle and crush easily.) Frozen herbs keep for about a year.

Drying ❧

Most herbs dry well with little loss of aroma and flavor. They may be dried outdoors, indoors, or in an oven.

If you dry them outdoors, place the herbs on a screen. Place the screen in a dry, shady area, never in direct sunlight. For best results bring the screen indoors overnight to avoid evening dampness and morning dew.

Herbs may be placed on a tray and dried indoors in any warm, dark, dry, well-ventilated area. They also may be tied in a bunch (stem ends up) and hung to dry in a kitchen. Whether they are dried indoors or outdoors, depending on the variety fresh herbs will dry in a week or two.

If you want to speed up the process, herbs may be dried in an oven. Arrange them on a cookie sheet and place the sheet in an oven set to about 100°F. Depending on the variety the herbs will dry in twenty-four to forty-eight hours.

Place dried herbs in a tightly sealed glass jar. Remember that a dry herb is more potent than a fresh herb. If a recipe calls for a tablespoon of a fresh herb, use a teaspoon or less of the dry herb in its stead.

Appendix

Sources for Recipes and Information

A lthough the following trade groups and organizations primarily serve the produce industry, they gladly furnish the general public with recipes and information. They try to answer any question posed by the consumer about their products.

APPLES

International Apple Institute
P.O. Box 1137
McLean, VA 22101

APRICOTS

Washington State Fruit Commission
1005 Tieton Drive
Yakima, WA 98902

ARTICHOKES

California Artichoke Advisory Board
P.O. Box 747
10719 Merritt Street
Castroville, CA 95012

AVOCADOS

California Avocado Commission
17620 Fitch
Irvine, CA 92714

Florida Lime and Avocado Administrative
Committee
P.O. Box 188
Homestead, FL 33030

BANANAS

Chiquita Brands, Inc.
15 Mercedes Drive
Montvale, NJ 07645

BLUEBERRIES

North American Blueberry Council
P.O. Box 166
Marmora, NJ 08223

BROCCOLI

D'Arrigo Bros. (Andy Boy)
P.O. Box 850
Salinas, CA 93901

CHERRIES

Washington State Fruit Commission
1005 Tieton Drive
Yakima, WA 98902

CITRUS FRUITS

Florida Dept. of Citrus
P.O. Box 148
Lakeland, FL 33802

Sunkist Growers, Inc.
P.O. Box 7888
Van Nuys, CA 91409

CORN

Florida Sweet Corn Exchange
P.O. Box 20155
Orlando, FL 32814

CRANBERRIES

Ocean Spray Cranberries, Inc.
Fresh Cranberry Division
Water Street
Plymouth, MA 02360

EXOTIC ITEMS

Frieda Produce Specialties
P.O. Box 58488
Los Angeles, CA 90058

GRAPES	California Table Grape Commission P.O. Box 5498 Fresno, CA 93755
KIWIS	Kiwi Growers of California 1151 Los Rios Drive Carmichael, CA 95608
	New Zealand Kiwi Fruit Authority 1714 Stockton Street San Francisco, CA 94111
LETTUCE	Leafy Greens Council 2 North Riverside Plaza Chicago, IL 60606
MUSHROOMS	American Mushroom Institute P.O. Box 373 Kennett Square, PA 19348
NECTARINES	California Tree Fruit Agreement P.O. Box 4640 Sacramento, CA 95865
ONIONS	National Onion Association 5701 East Evans Avenue Denver, CO 80222
PAPAYAS	Hawaiian Papaya Commission 55 Union Street San Francisco, CA 94111
PEACHES	National Peach Council P.O. Box 1085 Martinsburg, West VA 25401

PEARS   California Tree Fruit Agreement
                          P.O. Box 4640
                          Sacramento, CA 95865

PINEAPPLES   Castle and Cooke Foods
                          50 California Street
                          San Francisco, CA 94111

                          Del Monte Banana Co.
                          P.O. Box 11940
                          Miami, FL 33101

                          Pineapple Growers Assoc. of Hawaii
                          1902 Financial Plaza
                          Honolulu, Hawaii 96813

PLUMS   California Tree Fruit Agreement
                          P.O. Box 4640
                          Sacramento, CA 95865

POTATOES   Idaho Potato Commission
                          P.O. Box 1068
                          Boise, ID 83701

RUTABAGAS   Ontario Rutabaga Marketing Board
                          P.O. Box 328
                          150 Alice Street
                          Lucan, Ontario, Canada N0M 2J0

STRAWBERRIES   California Strawberry Advisory Board
                          P.O. Box 269
                          Watsonville, CA 95077

SWEET POTATOES   North Carolina Yam Commission
                          P.O. Box 12005
                          Raleigh, NC 27605

WATERMELONS

National Watermelon Association
P.O. Box 38
Morven, GA 31638

GENERAL INFORMATION

United Fresh Fruit and Vegetable
Association
North Washington and Madison Avenues
Alexandria, VA 22314

or

Write your own State Department of
Agriculture or the Cooperative Extension
of any agricultural college

The source of most of the above addresses is *The Packer,* published in Shawnee Mission, KS. It is the largest trade newspaper of the produce industry.

Index